Encyclopaedia of

Tools and Techniques in Biochemistry and Molecular Biology

Encyclopaedia of

Tools and Techniques in Biochemistry and Molecular Biology

Volume 3

Editor:

Rashmi Rana

AGROTECH PRESS

JAIPUR ▪ NEW DELHI

AGROTECH PRESS

Regd. Off. "Bony Residency", Gate No. 2
Opp. Tilak Public School, Vishweswariya Nagar,
Gopalpura Road, Jaipur - 302018, India
Ph: 0141 - 2761280, Telefax: 2761381
E-mail : agrotechpress@gmail.com

Delhi Office: 102, 1st Floor, Satyam House,
4327/3, Ansari Road, Darya Ganj,
New Delhi - 110002, India
Ph: 011-45652440
visit us: www.agrotechpress.com

Distributed By
KEYA INTERNATIONAL PRESS LLP
E-mail: keyainternationalpressllp@gmail.com

Published in India, 2025

ISBN : 9789384568245 (4 Vols. Set)

Typeset by Thomson Press

Printed and bound in India by Thomson Press Ltd.,
New Delhi

Contents

Preface

Biochemistry is the study of the structure, composition, and chemical reactions of substances in living systems. Biochemistry emerged as a separate discipline when scientists combined biology with organic, inorganic, or physical chemistry and began to study such topics as how living things obtain energy from food, the chemical basis of heredity, and what fundamental changes occur in disease. Biochemistry includes the sciences of molecular biology; immunochemistry; neurochemistry; and bioinorganic, bioorganic, and biophysical chemistry. Biochemistry is applied to medicine, dentistry, and veterinary medicine. In food science, biochemists research ways to develop abundant and inexpensive sources of nutritious foods, determine the chemical composition of foods, develop methods to extract nutrients from waste products, or invent ways to prolong the shelf life food products. In agriculture, biochemists study the interaction of herbicides with plants. They examine the structure-activity relationships of compounds, determine their ability to inhibit growth, and evaluate the toxicological effects on surrounding life. Biochemistry spills over into pharmacology, physiology, microbiology, and clinical chemistry. In these areas, a biochemist may investigate the mechanism of a drug action; engage in viral research; conduct research pertaining to organ function; or use chemical concepts, procedures, and techniques to study the diagnosis and therapy of disease and the assessment of health. Work in the field of biochemistry is often related to toxicology. Rogene Henderson, senior scientist and supervisor of the Biochemical Toxicology Group at Lovelace Respiratory Research Institute, does research to understand ways in which organic compounds in the body are changed by enzymes into toxic metabolites. Henderson focuses on determining the health effects of inhaled pollutants. She develops chemical analytical techniques to detect pollutants and their metabolites in body tissues and fluids, uses mathematics to describe the relationships between the air and body concentrations of these chemicals or their metabolites, and determines how these concentrations change with time.

Genomics is a discipline in genetics concerned with the study of the genomes of organisms. The field includes efforts to determine the entire DNA sequence of organisms and fine-scale genetic mapping. The field also includes studies of intragenomic phenomena such as heterosis, epistasis, pleiotropy and other interactions between loci and alleles within the genome. In contrast, the investigation of the roles and functions of single genes is a primary focus of molecular biology or genetics and is a common topic of modern medical and biological research. Research of single genes does not fall into the definition of genomics unless the aim of this genetic, pathway, and functional information analysis is to elucidate its effect on, place in, and response to the entire genome's networks. For the United States Environmental Protection Agency, "the term 'genomics' encompasses a broader scope of scientific inquiry associated technologies than when genomics was initially considered. A genome is the sum total of all an individual organism's genes. Thus, genomics is the study of all the genes of a cell, or tissue, at the DNA, mRNA, or protein levels." The first genomes to be sequenced were those of a virus and a mitochondrion, and were done by Fred Sanger. His group established techniques of sequencing, genome mapping, data storage, and bioinformatic analyses in the 1970-1980s. A major branch of genomics is still concerned with sequencing the genomes of various organisms, but the knowledge of full genomes has created the possibility for the field of functional genomics, mainly concerned with patterns of gene expression during various conditions. The most important tools here are microarrays and bioinformatics. Study of the full set of proteins in a cell type or tissue, and the changes during various conditions, is called proteomics. A related concept is materiomics, which is defined as the holistic study of the material properties of biological materials, and their effect on the macroscopic function and failure in their biological context. The actual term 'genomics' is thought to have been coined by Dr. Tom Roderick, a geneticist at the Jackson Laboratory over beer at a meeting held in Maryland on the mapping of the human genome in 1986.

The purpose of this book is to provide a deeper insight into the chemistry behind genomics revolution. It takes readers on a fascinating journey through genomics-from the basic presentation of ideas about heredity through the essential principles of molecular biology. Written for anyone wondering why we are the way we are, this book is a timely and important source of information that spans the science, the promise, and the potential pitfalls of a field moving so quickly that its achievements may alternately delight and trouble us, but almost always surprise us.

—Editor

1

Analysing Nucleic Acids

In molecular biology, analysis of nucleic acids is commonly performed to determine the average concentrations of DNA or RNA present in a mixture, as well as their purity. Reactions that use nucleic acids often require particular amounts and purity for optimum performance. There are several methods to establish the concentration of a solution of nucleic acids, including spectrophotometric quantification and UV fluorescence in presence of a DNA dye.

Spectrophotometric Analysis

Nucleic acids absorb ultraviolet light in a specific pattern. In a spectrophotometer, a sample is exposed to ultraviolet light at 260 nm, and a photo-detector measures the light that passes through the sample. The more light absorbed by the sample, the higher the nucleic acid concentration in the sample.

Using the Beer Lambert Law it is possible to relate the amount of light absorbed to the concentration of the absorbing molecule. At a wavelength of 260 nm, the average extinction coefficient for double-stranded DNA is 0.020 $(\mu g/ml)^{-1}\ cm^{-1}$, for single-stranded DNA and RNA it is 0.027 $(\mu g/ml)^{-1}\ cm^{-1}$ and for short single-stranded oligonucleotides it is dependent on the length and base composition. Thus, an optical density (or "OD") of 1 corresponds to a concentration of 50 μg/ml for double-stranded DNA. This method of calculation is valid for up to an OD of at least 2. A more accurate extinction coefficient may be needed for oligonucleotides; these can be predicted using the nearest-neighbor model.

Sample Purity

It is common for nucleic acid samples to be contaminated with other molecules (i.e. proteins, organic compounds, other). The ratio of the absorbance at 260 and 280nm ($A_{260/280}$) is used to assess the purity of nucleic acids. For pure DNA, $A_{260/280}$ is ~1.8 and for pure RNA $A_{260/280}$ is ~2.

Protein Contamination and the 260:280 Ratio

The ratio of absorptions at 260nm vs 280nm is commonly used to assess DNA contamination of protein solutions, since proteins (in particular, the aromatic amino acids) absorb light at 280nm. The reverse, however, is not true — it takes a relatively large amount of protein contamination to significantly affect the 260:280 ratio in a nucleic acid solution. 260:280 ratio has high sensitivity for nucleic acid contamination in protein:

% protein	*% nucleic acid*	*260:280 ratio*
100	0	0.57
95	5	1.06
90	10	1.32
70	30	1.73

260:280 ratio lacks sensitivity for protein contamination in nucleic acids:

% nucleic acid	*% protein*	*260:280 ratio*
100	0	2.00
95	5	1.99
90	10	1.98
70	30	1.94

This difference is due to the much higher extinction coefficient nucleic acids have at 260nm and 280nm, compared to that of proteins. Because of this, even for relatively high concentrations of protein, the protein contributes relatively little to the 260 and 280 absorbance. While the protein contamination cannot be reliably assessed with a 260:280 ratio, this also means that it contributes little error to DNA quantity estimation.

Other Common Contaminants

- Contamination by phenol, which is commonly used in nucleic acid purification, can significantly throw off quantification estimates. Phenol absorbs with a peak at 270nm and a $A_{260/280}$ of 2. Nucleic acid preparitions uncontaminated by phenol should have a $A_{260/270}$ of around 1.2. Contamination by phenol can significantly contribute to overestimation of DNA concentration.

- Absorption at 230nm can be caused by contamination by phenolate ion, thiocyanates, and other organic compounds. For a pure RNA sample, the $A_{260/230}$ should be around 2, and for a pure DNA sample, the $A_{260/230}$ should be around 1.8.
- Absorption at 330nm and higher indicates particulates contaminating the solution, causing scattering of light in the visible range. The value in a pure nucleic acid sample should be zero.
- Negative values could result if an incorrect solution was used as blank. Alternatively, these values could arise due to fluorescence of a dye in the solution.

Quantification using Fluorescent Dyes

An alternative way to assess DNA concentration is to use measure the fluorescence intensity of dyes that bind to nucleic acids and selectively fluoresce when bound (eg. Ethidium bromide). This method is useful for cases where concentration is too low to accurately assess with spectrophotometry and in cases where contaminants absorbing at 260nm make accurate quantitation by that method impossible.

There are two main ways to approach this. "Spotting" involves placing a sample directly onto an agarose gel or plastic wrap. The fluorescent dye is either present in the agarose gel, or is added in appropriate concentrations to the samples on the plastic film. A set of samples with known concentrations are spotted alongside the sample. The concentration of the unknown sample is then estimated by comparison with the fluorescence of these known concentrations. Alternatively, one may run the sample through an agarose or polyacrylamide gel, alongside some samples of known concentration. As with the spot test, concentration is estimated through comparison of fluorescent intensity with the known samples. If the sample volumes are large enough to use microplates or cuvettes, the dye-loaded samples can also be quantified with a fluorescence photometer.

Nucleic Acid Methods

Nucleic acid methods are the techniques used to study nucleic acids (DNA and RNA).

Purification

Phenol-chloroform Extraction

Phenol-chloroform extraction is a liquid-liquid extraction technique in biochemistry. It is widely used in molecular biology for isolating DNA, RNA and protein. Equal volumes of a phenol:chloroform mixture and an aqueous sample are mixed, forming a biphasic mixture. This method may take longer than a column-based system such as the silica-based purification, but has higher purity and the advantage of high recovery of RNA: an RNA column is typically unsuitable for purification of short (<200 nucleotides) RNA species, such as siRNA, miRNA and tRNA. Column methods also shear large DNA fragments, which may or may not be a problem depending on downstream applications. It was originally devised by Piotr Chomczynski and Nicoletta Sacchi and published in 1987 (referred to as *Guanidinium thiocyanate-phenol-chloroform extraction*). It is sold by Sigma-Aldrich by the name TRI Reagent, by Invitrogen under the name TRIzol and by Bioline as Trisure.

How it Works

This method relies on phase separation by centrifugation of a mix of the aqueous sample and a solution containing water-saturated phenol, chloroform and a chaotropic denaturing solution (guanidinium thiocyanate) resulting in an upper aqueous phase and a lower organic phase (mainly chloroform). Nearly all of the RNA is present in the aqueous phase, while DNA and protein partition in the interphase and organic phase, respectively. In a last step, RNA is recovered from the aqueous phase by precipitation with 2-propanol or ethanol. DNA will be located in the aqueous phase in the absence of guanidinium thiocyanate and thus the technique can be used for DNA purification alone.

Guanidinium thiocyanate denatures proteins, including RNases, and separates rRNA from ribosomes, while phenol, isopropanol and water are solvents with poor solubility. In the presence of chloroform or BCP (bromochloropropane), these solvents separate entirely into two phases that are recognized by their colour: a clear, upper aqueous phase (containing the nucleic acids) and a bright pink lower phase (containing the proteins dissolved in phenol and the lipids dissolved in chloroform). Other denaturing chemicals such as 2-mercaptoethanol and sarcosine may also be used. The major downside is that Phenol and chloroform are both hazardous and inconvenient materials, and the extraction is often more laborious, so in recent years many companies now offer alternative ways to isolate DNA.

Reagents

- *Phenol*: The phenol used for biochemistry comes as a water-saturated solution with Tris buffer, as a Tris-buffered 50% phenol, 50% chloroform solution, or as a Tris-buffered 50% phenol, 48% chloroform, 2% isoamyl alcohol solution (sometimes called "25:24:1"). Phenol is naturally somewhat water-soluble, and gives a fuzzy interface, which is sharpened by the presence of chloroform, and the isoamyl alcohol reduces foaming. Most solutions also have an antioxidant, as oxidized phenol damages the nucleic acids. For RNA purification, the pH is kept around pH 4, which retains RNA in the aqueous phase preferentially. For DNA purification, the pH is usually near 7, at which point all nucleic acids are found in the aqueous phase.

- *Chloroform*: Chloroform is stabilized with small quantities of amylene or ethanol, because exposure of pure chloroform to oxygen and ultraviolet light produces phosgene gas. Some chloroform solutions come as pre-made a 96% chloroform, 4% isoamyl alcohol mixtures that can be mixed with an equal volume of phenol to obtain the 25:24:1 solution.
- *Isoamyl alcohol*: Some protocols include isoamyl alcohol as a stabilizing agent, while others do not require it at all.
- Minicolumn purification: Column-based nucleic acid purification is a solid phase extraction method to quickly purify nucleic acids.

This method relies on the fact that the nucleic acid may bind (adsorption) to the solid phase (silica or other) depending on the pH and the salt content of the buffer, which may be a Tris-EDTA (TE) buffer or Phosphate buffer (used in DNA microarray experiments due to the reactive amines).

Therefore, three stages are:

- The sample is added to the column and the nucleic acid binds thanks to the lower pH (relative to the silanol groups on the column) and salt concentration of the binding solution, which may contain buffer, a denaturing agent (such as guanidine hydrochloride), Triton X-100, isopropanol and a pH indicator
- The column is then washed (5 mM KPO4 pH 8.0 or similar, 80% EtOH)
- The column can be eluted with buffer or simply water.

Even prior to the major techniques employed today it was known that in the presence of chaotropic agents, such as sodium iodide or sodium perchlorate, DNA binds to silica, glass particles or to unicellular algae called diatoms which shield their cell walls with silica. This propriety was used to purify nucleic acid using glass powder or silica beads under alkaline conditions. This was later improved used guanidinium thiocyanate or guanidinium hydrochloride as the chaotropic agent. The use of beads was later changed to minicolumns.

Quantification:

- Abundance in weight: spectroscopic quantification
- Absolute abundance in number: Q-PCR
- high-throughput relative abundance DNA microarray
- high-throughput absolute abundance SAGE
- Size: Gel electrophoresis.

Synthesis:

- De novo: Oligonucleotide synthesis
- Amplification: PCR.

Other:

- Nucleic acid simulations
- DNA sequencing.

The term DNA sequencing refers to sequencing methods for determining the order of the nucleotide bases—adenine, guanine, cytosine, and thymine—in a molecule of DNA. Knowledge of DNA sequences has become indispensable for basic biological research, other research branches utilizing DNA sequencing, and in numerous applied fields such as diagnostic, biotechnology, forensic biology and biological systematics. The advent of DNA sequencing has significantly accelerated biological research and discovery. The rapid speed of sequencing attained with modern DNA sequencing technology has been instrumental in the sequencing of the human genome, in the Human Genome Project. Related projects, often by scientific collaboration across continents, have generated the complete DNA sequences of many animal, plant, and microbial genomes.

The first DNA sequences were obtained in the early 1970s by academic researchers using laborious methods based on two-dimensional chromatography. Following the development of dye-based sequencing methods with automated analysis, DNA sequencing has become easier and orders of magnitude faster.

History

RNA sequencing was one of the earliest forms of nucleotide sequencing. The major landmark of RNA sequencing is the sequence of the first complete gene and the complete genome of Bacteriophage MS2, identified and published by Walter Fiers and his coworkers at the University of Ghent (Ghent, Belgium), between 1972 and 1976. Prior to the development of rapid DNA sequencing methods in the early 1970s by Frederick Sanger at the University of Cambridge, in England and Walter Gilbert and Allan Maxam at Harvard, a number of laborious methods were used. For instance, in 1973, Gilbert and Maxam reported the sequence of 24 basepairs using a method known as wandering-spot analysis.

The chain-termination method developed by Sanger and coworkers in 1975 soon became the method of choice, owing to its relative ease and reliability.

Maxam–Gilbert Sequencing

In 1976–1977, Allan Maxam and Walter Gilbert developed a DNA sequencing method based on chemical modification of DNA and subsequent cleavage at specific bases. Although Maxam and Gilbert published their chemical sequencing method two years after the ground-breaking paper of Sanger and

Coulson on plus-minus sequencing, Maxam–Gilbert sequencing rapidly became more popular, since purified DNA could be used directly, while the initial Sanger method required that each read start be cloned for production of single-stranded DNA. However, with the improvement of the chain-termination method, Maxam-Gilbert sequencing has fallen out of favour due to its technical complexity prohibiting its use in standard molecular biology kits, extensive use of hazardous chemicals, and difficulties with scale-up.

The newly synthesized and labelled DNA fragments are heat denatured, and separated by size (with a resolution of just one nucleotide) by gel electrophoresis on a denaturing polyacrylamide-urea gel with each of the four reactions run in one of four individual lanes (lanes A, T, G, C); the DNA bands are then visualized by autoradiography or UV light, and the DNA sequence can be directly read off the X-ray film or gel image. In the image on the right, X-ray film was exposed to the gel, and the dark bands correspond to DNA fragments of different lengths. A dark band in a lane indicates a DNA fragment that is the result of chain termination after incorporation of a dideoxynucleotide (ddATP, ddGTP, ddCTP, or ddTTP). The relative positions of the different bands among the four lanes are then used to read (from bottom to top) the DNA sequence.

Technical variations of chain-termination sequencing include tagging with nucleotides containing radioactive phosphorus for radiolabelling, or using a primer labelled at the 5' end with a fluorescent dye. Dye-primer sequencing facilitates reading in an optical system for faster and more economical analysis and automation. The later development by Leroy Hood and coworkers of fluorescently labelled ddNTPs and primers set the stage for automated, high-throughput DNA sequencing. Chain-termination methods have greatly simplified DNA sequencing. For example, chain-termination-based kits are commercially available that contain the reagents needed for sequencing, pre-aliquoted and ready to use. Limitations include non-specific binding of the primer to the DNA, affecting accurate read-out of the DNA sequence, and DNA secondary structures affecting the fidelity of the sequence.

Dye-terminator Sequencing

Dye-terminator sequencing utilizes labelling of the chain terminator ddNTPs, which permits sequencing in a single reaction, rather than four reactions as in the labelled-primer method. In dye-terminator sequencing, each of the four dideoxynucleotide chain terminators is labelled with fluorescent dyes, each of which emit light at different wavelengths. Owing to its greater expediency and speed, dye-terminator sequencing is now the mainstay in automated sequencing. Its limitations include dye effects due to differences in the incorporation of the dye-labelled chain terminators into the DNA fragment, resulting in unequal peak heights and shapes in the electronic DNA sequence trace chromatogram after capillary electrophoresis.

This problem has been addressed with the use of modified DNA polymerase enzyme systems and dyes that minimize incorporation variability, as well as methods for eliminating "dye blobs". The dye-terminator sequencing method, along with automated high-throughput DNA sequence analyzers, is now being used for the vast majority of sequencing projects.

Challenges

Common challenges of DNA sequencing include poor quality in the first 15–40 bases of the sequence and deteriorating quality of sequencing traces after 700–900 bases. Base calling software typically gives an estimate of quality to aid in quality trimming. In cases where DNA fragments are cloned before sequencing, the resulting sequence may contain parts of the cloning vector.

In contrast, PCR-based cloning and emerging sequencing technologies based on pyrosequencing often avoid using cloning vectors. Recently, one-step Sanger sequencing (combined amplification and sequencing) methods such as Ampliseq and SeqSharp have been developed that allow rapid sequencing of target genes without cloning or prior amplification.

Current methods can directly sequence only relatively short (300–1000 nucleotides long) DNA fragments in a single reaction. The main obstacle to sequencing DNA fragments above this size limit is insufficient power of separation for resolving large DNA fragments that differ in length by only one nucleotide.

Automation and Sample Preparation

Automated DNA-sequencing instruments (DNA sequencers) can sequence up to 384 DNA samples in a single batch (run) in up to 24 runs a day. DNA sequencers carry out capillary electrophoresis for size separation, detection and recording of dye fluorescence, and data output as fluorescent peak trace chromatograms. Sequencing reactions by thermocycling, cleanup and re-suspension in a buffer solution before loading onto the sequencer are performed separately. A number of commercial and non-commercial software packages can trim low-quality DNA traces automatically. These programs score the quality of each peak and remove low-quality base peaks (generally located at the ends of the sequence). The accuracy of such algorithms is below visual examination by a human operator, but sufficient for automated processing of large sequence data sets.

Amplification and Clonal Selection

Large-scale sequencing aims at sequencing very long DNA pieces, such as whole chromosomes. Common approaches consist of cutting (with restriction enzymes) or shearing (with mechanical forces) large DNA fragments into shorter DNA fragments. The fragmented DNA is cloned into a DNA vector, and amplified in *Escherichia coli*. Short DNA fragments purified from individual bacterial colonies are individually sequenced and assembled electronically into one long, contiguous sequence.

This method does not require any pre-existing information about the sequence of the DNA and is referred to as *de novo* sequencing. Gaps in the assembled sequence may be filled by primer walking. The different strategies have different tradeoffs in speed and accuracy; *shotgun methods* are often used for sequencing large genomes, but its assembly is complex and difficult, particularly with sequence repeats often causing gaps in genome assembly.

Most sequencing approaches use an *in vitro* cloning step to amplify individual DNA molecules, because their molecular detection methods are not sensitive enough for single molecule sequencing. Emulsion PCR isolates individual DNA molecules along with primer-coated beads in aqueous droplets within an oil phase.

Polymerase chain reaction (PCR) then coats each bead with clonal copies of the DNA molecule followed by immobilization for later sequencing. Emulsion PCR is used in the methods by Marguilis et al. (commercialized by 454 Life Sciences), Shendure and Porreca et al. (also known as "Polony sequencing") and SOLiD sequencing, (developed by Agencourt, now Applied Biosystems).

Another method for *in vitro* clonal amplification is *bridge PCR*, where fragments are amplified upon primers attached to a solid surface, used in the Illumina Genome Analyzer. The single-molecule method developed by Stephen Quake's laboratory (later commercialized by Helicos) is an exception: it uses bright fluorophores and laser excitation to detect pyrosequencing events from individual DNA molecules fixed to a surface, eliminating the need for molecular amplification.

High-throughput Sequencing

The high demand for low-cost sequencing has driven the development of high-throughput sequencing technologies that parallelize the sequencing process, producing thousands or millions of sequences at once. High-throughput sequencing technologies are intended to lower the cost of DNA sequencing beyond what is possible with standard dye-terminator methods.

Lynx Therapeutics' Massively Parallel Signature Sequencing (MPSS)

The first of the "next-generation" sequencing technologies, MPSS was developed in 1990s at Lynx Therapeutics, a company founded in 1992 by Sidney Brenner and Sam Eletr. MPSS was a bead-based method that used a complex approach of adapter ligation followed by adapter decoding, reading the sequence in increments of four nucleotides; this method made it susceptible to sequence-specific bias or loss of specific sequences.

Because the technology was so complex, MPSS was only performed 'in-house' by Lynx Therapeutics and no machines were sold; when the merger with Solexa later lead to the development of sequencing-by-synthesis, a more simple approach with numerous advantages, MPSS became obsolete. However, the essential properties of the MPSS output were typical of later "next-gen" data types, including hundreds of thousands of short DNA sequences. In the case of MPSS, these were typically used for sequencing cDNA for measurements of gene expression levels. Lynx Therapeutics merged with Solexa in 2004, and this company was later purchased by Illumina.

454 Pyrosequencing

A parallelized version of pyrosequencing was developed by 454 Life Sciences. The method amplifies DNA inside water droplets in an oil solution (emulsion PCR), with each droplet containing a single DNA template attached to a single primer-coated bead that then forms a clonal colony. The sequencing machine contains many picolitre-volume wells each containing a single bead and sequencing enzymes.

Pyrosequencing uses luciferase to generate light for detection of the individual nucleotides added to the nascent DNA, and the combined data are used to generate sequence read-outs. This technology provides intermediate read length and price per base compared to Sanger sequencing on one end and Solexa and SOLiD on the other. 454 Life Sciences has since been acquired by Roche Diagnostics.

Illumina (Solexa) Sequencing

Solexa, now part of Illumina developed a sequencing technology based on reversible dye-terminators. DNA molecules are first attached to primers on a slide and amplified so that local clonal colonies are formed (bridge amplification). One type of nucleotide at a time is then added, and non-incorporated nucleotides are washed away. Unlike pyrosequencing, the DNA can only be extended one nucleotide at a time. A camera takes images of the fluorescently labelled nucleotides and the dye is chemically removed from the DNA, allowing a next cycle.

SOLiD Sequencing

Applied Biosystems' SOLiD technology employs sequencing by ligation. Here, a pool of all possible oligonucleotides of a fixed length are labelled according to the sequenced position. Oligonucleotides are annealed and ligated; the preferential ligation by DNA ligase for matching sequences results in a signal informative of the nucleotide at that position.

Before sequencing, the DNA is amplified by emulsion PCR. The resulting bead, each containing only copies of the same DNA molecule, are deposited on a glass slide. The result is sequences of quantities and lengths comparable to Illumina sequencing.

Future Methods

Sequencing by hybridization is a non-enzymatic method that uses a DNA microarray. A single pool of DNA whose sequence is to be determined is fluorescently labelled and hybridized to an array containing known sequences. Strong hybridization signals from a given spot on the array identifies its sequence in the DNA being sequenced. Mass spectrometry may be used to determine mass differences between DNA fragments produced in chain-termination reactions.

DNA sequencing methods currently under development include labelling the DNA polymerase, reading the sequence as a DNA strand transits through nanopores, and microscopy-based techniques, such as AFM or electron microscopy that are used to identify the positions of individual nucleotides within long DNA fragments (>5,000 bp) by nucleotide labelling with heavier elements (e.g., halogens) for visual detection and recording.

In microfluidic Sanger sequencing the entire thermocycling amplification of DNA fragments as well as their separation by electrophoresis is done on a single glass wafer (approximately 10 cm in diameter) thus reducing the reagent usage as well as cost. In some instances researchers have shown that they can increase the throughput of conventional sequencing through the use of microchips.

Research will still need to be done in order to make this use of technology effective. In October 2006, the X Prize Foundation established an initiative to promote the development of full genome sequencing technologies, called the Archon X Prize, intending to award $10 million to "the first Team that can build a device and use it to sequence 100 human genomes within 10 days or less, with an accuracy of no more than one error in every 100,000 bases sequenced, with sequences accurately covering at least 98% of the genome, and at a recurring cost of no more than $10,000 (US) per genome."

Bisulfite Sequencing

Bisulfite sequencing is the use of bisulfite treatment of DNA to determine its pattern of methylation. DNA methylation was the first discovered epigenetic mark, and remains the most studied. In animals it predominantly involves the addition of a methyl group to the carbon-5 position of cytosine residues of the dinucleotide CpG, and is implicated in repression of transcriptional activity.

Treatment of DNA with bisulfite converts cytosine residues to uracil, but leaves 5-methylcytosine residues unaffected. Thus, bisulfite treatment introduces specific changes in the DNA sequence that depend on the methylation status of individual cytosine residues, yielding single- nucleotide resolution information about the methylation status of a segment of DNA. Various analyses can be performed on the altered sequence to retrieve this information. The objective of this analysis is therefore reduced to differentiating between single nucleotide polymorphisms (cytosines and thymidine) resulting from bisulfite conversion.

Methods

Bisulfite sequencing applies routine sequencing methods on bisulfite-treated genomic DNA to determine methylation status at CpG dinucleotides. Other non-sequencing strategies are also employed to interrogate the methylation at specific loci or at a genome-wide level. All strategies assume that bisulfite-induced conversion of unmethylated cytosines to uracil is complete, and this serves as the basis of all subsequent techniques. Ideally, the method used would determine the methylation status separately for each allele. Alternative methods to bisulfite sequencing include methylation-specific restriction analysis and methylated DNA immunoprecipitation (MeDIP).

Methodologies to analyze bisulfite-treated DNA are continuously being developed. To summarize these rapidly evolving methologies, numerous review articles have been written. The methodologies can be generally divided into strategies based on methylation-specific PCR (MSP), and strategies employing polymerase chain reaction (PCR) performed under non-methylation-specific conditions. Microarray-based methods use PCR based on non-methylation-specific conditions also.

Non-methylation-specific PCR based Methods

Direct Sequencing

The first reported method of methylation analysis using bisulfite-treated DNA utilized PCR and standard dideoxynucleotide DNA sequencing to directly determine the nucleotides resistant to bisulfite conversion. Primers are designed to be strand-specific as well as bisulfite-specific (i.e., primers containing non-CpG cytosines such that they are not complementary to non-bisulfite-treated DNA), flanking (but not involving) the methylation site of interest.

Therefore, it will amplify both methylated and unmethylated sequences, in contrast to methylation-specific PCR. All sites of unmethylated cytosines are displayed as thymines in the resulting amplified sequence of the sense strand, and as adenines in the amplified antisense strand. This technique required cloning of the PCR product prior to sequencing for adequate sensitivity, and therefore was a very labour-intensive method unsuitable for higher throughput. Alternatively, nested PCR methods can be used to enhance the product for sequencing.

All subsequent DNA methylation analysis techniques using bisulfite-treated DNA is based on this report by Frommer et al.. Although most other modalities are not true sequencing-based techniques, the term "bisulfite sequencing" is often used to describe bisulfite-conversion DNA methylation analysis techniques in general.

Pyrosequencing

Pyrosequencing has also been used to analyze bisulfite-treated DNA without using methylation-specific PCR. Following PCR amplification of the region of interest, Pyrosequencing is used to determine the bisulfite-converted sequence of specific CpG sites in the region. The ratio of C-to-T at individual sites can be determined quantitatively based on the amount of C and T incorporation during the sequence extension. The main limitation of this method is the cost of the technology. However, Pyrosequencing does well allow for extension to high-throughput screening methods.

A further improvement to this technique was recently described by Wong et al., which uses allele-specific primers that incorporate single-nucleotide polymorphisms into the sequence of the sequencing primer, thus allowing for separate analysis of maternal and paternal alleles. This technique is of particular usefulness for genomic imprinting analysis.

Methylation-sensitive Single-strand Conformation analysis (MS-SSCA)

This method is based on the single-strand conformation polymorphism analysis (SSCA) method developed for single-nucleotide polymorphism (SNP) analysis. SSCA differentiates between single-stranded DNA fragments of identical size but distinct sequence based on differential migration in non-denaturating electrophoresis. In MS-SSCA, this is used to distinguish between bisulfite-treated, PCR-amplified regions containing the CpG sites of interest. Although SSCA lacks sensitivity when only a single nucleotide difference is present, bisulfite treatment frequently makes a number of C-to-T conversions in most regions of interest, and the resulting sensitivity approaches 100%. MS-SSCA also provides semi-quantitative analysis of the degree of DNA methylation based on the ratio of band intensities. However, this method is designed to assess all CpG sites as a whole in the region of interest rather than individual methylation sites.

High Resolution Melting Analysis (HRM)

A further method to differentiate converted from unconverted bisulfite-treated DNA is using high-resolution melting analysis (HRM), a real-time PCR-based technique initially designed to distinguish SNPs. The PCR amplicons are analysed directly by temperature ramping and resulting liberation of an intercalating fluorescent dye during melting. The degree of methylation, as represented by the C-to-T content in the amplicon, determines the rapidity of melting and consequent release of the dye. This method allows direct quantitation in a single-tube assay, but assesses methylation in the amplified region as a whole rather than at specific CpG sites.

Methylation-sensitive Single-nucleotide Primer Extension (MS-SnuPE)

MS-SnuPE employs the primer extension method initially designed for analysing single-nucleotide polymorphisms. DNA is bisulfite-converted, and bisulfite-specific primers are annealed to the sequence up to the base pair immediately before the CpG of interest. The primer is allowed to extend one base pair into the C (or T) using DNA polymerase terminating dideoxynucleotides, and the ratio of C to T is determined quantitatively.

A number of methods can be used to determine this C:T ratio. At the beginning, MS-SnuPE relied on radioactive ddNTPs as the reporter of the primer extension. Fluorescence-based methods or Pyrosequencing can also be used. However, matrix-assisted laser desorption ionization/time-of-flight (MALDI-TOF) mass spectrometry analysis to differentiate between the two polymorphic primer extension products can be used, in essence, based on the GOOD assay designed for SNP genotyping. Ion pair reverse-phase high-performance liquid chromatography (IP-RP-HPLC) has also been used to distinguish primer extension products.

Base-specific Cleavage/MALDI-TOF

A recently described method by Ehrich et al. further takes advantage of bisulfite-conversions by adding a base-specific cleavage step to enhance the information gained from the nucleotide changes. By first using in vitro transcription of the region of interest into RNA (by adding an RNA polymerase promoter site to the PCR primer in the initial amplification), RNase A can be used to cleave the RNA transcript at base-specific sites. As RNase A cleaves RNA specifically at cytosine and uracil ribonucleotides, base-specificity is achieved by adding incorporating cleavage-resistant dTTP when cytosine-specific (C-specific) cleavage is desired, and incorporating dCTP when uracil-specific (U-specific) cleavage is desired. The cleaved fragments can then be analysed by MALDI-TOF. Bisulfite treatment results in either introduction/ removal of cleavage sites by C-to-U conversions or shift in fragment mass by G-to-A conversions in the amplified reverse strand. C-specific cleavage will cut specifically at all methylated CpG sites. By analysing the sizes of the resulting fragments, it is possible to determine the specific pattern of DNA methylation of CpG sites within the region, rather than determining the extent of methylation of the region as a whole. This method demonstrated efficacy for high-throughput screening, allowing for interrogation of numerous CpG sites in multiple tissues in a cost-efficient manner.

Methylation-specific PCR (MSP)

This alternative method of methylation analysis also uses bisulfite-treated DNA but avoids the need to sequence the area of interest. Instead, primer pairs are designed themselves to be "methylated-specific" by including sequences complementing only unconverted 5-methylcytosines, or, on the converse, "unmethylated-specific", complementing thymines converted from unmethylated cytosines.

Methylation is determined by the ability of the specific primer to achieve amplification. This method is particularly useful to interrogate CpG islands with possibly high methylation density, as increased numbers of CpG pairs in the primer increase the specificity of the assay. Placing the CpG pair at the 3'-end of the primer also improves the sensitivity. The initial report using MSP described sufficient sensitivity to detect methylation of 0.1% of alleles. In general, MSP and its related protocols are considered to be the most sensitive when interrogating the methylation status at a specific locus.

The MethyLight method is based on MSP, but provides a quantitative analysis using real-time PCR. Methylated-specific primers are used, and a methylated-specific fluorescence reporter probe is also used that anneals to the amplified region. In alternative fashion, the primers or probe can be designed without methylation specificity if discrimination is needed between the CpG pairs within the involved sequences. Quantitation is made in reference to a methylated reference DNA. A modification to this protocol to increase the specificity of the PCR for successfully bisulfite-converted DNA (ConLight-MSP) uses an additional probe to bisulfite-unconverted DNA to quantify this non-specific amplification.

Further methodology using MSP-amplified DNA analyzes the products using melting-curve analysis (Mc-MSP). This method amplifies bisulfite-converted DNA with both methylated-specific and unmethylated-specific primers, and determines the quantitative ratio of the two products by comparing the differential peaks generated in a melting-curve analysis. A high-resolution melting analysis method that uses both real-time quantification and melting analysis has been introduced, in particular, for sensitive detection of low-level methylation

Microarray-based Methods

Microarray-based methods are a logical extension of the technologies available to analyze bisulfite-treated DNA to allow for genome-wide analysis of methylation. Oligonucleotide microarrays are designed using oligonucleotide pairs targeting CpG sites of interest, with one complementary to the unaltered methylated sequence, and the other to the C-to-U-converted unmethylated sequence. The oligonucleotides are also bisulfite-specific to prevent binding to DNA incompletely converted by bisulfite. The Illumina Methylation Assay is one such assay that applies the bisulfite sequencing technology on a microarray level to generate genome-wide methylation data.

Limitations of Incomplete Conversion

Bisulphite sequencing relies on the conversion of every single unmethylated cytosine residue to uracil. If conversion is incomplete, the subsequent analysis will incorrectly interpret the unconverted unmethylated cytosines as methylated cytosines, resulting in false positive results for methylation. Only cytosines in single-stranded DNA are susceptible to attack by bisulphite, therefore denaturation of the DNA undergoing analysis is critical. It is important to ensure that reaction parameters such as temperature and salt concentration are suitable to maintain the DNA in a single-stranded conformation and allow for complete conversion. Embedding the DNA in agarose gel has been reported to improve the rate of conversion by keeping strands of DNA physically separate.

Degradation of DNA during Bisulphite Treatment

A major challenge in bisulphite sequencing is the degradation of DNA that takes place concurrently with the conversion. The conditions necessary for complete conversion, such as long incubation times, elevated temperature, and high bisulphite concentration, can lead to the degradation of about 90% of the incubated DNA. Given that the starting amount of DNA is often limited, such extensive degradation can be problematic. The degradation occurs as depurinations resulting in random strand breaks. Therefore the longer the desired PCR amplicon, the more limited the number of intact template molecules will likely be. This could lead to the failure of the PCR amplification, or the loss of quantitatively accurate information on methylation levels resulting from the limited sampling of template molecules. Thus, it is important to assess the amount of DNA degradation resulting from the reaction conditions employed, and consider how this will affect the desired amplicon. Techniques can also be used to minimize DNA degradation, such as cycling the incubation temperature.

Other Concerns

A potentially significant problem following bisulphite treatment is incomplete desulfonation of pyrimidine residues due to inadequate alkalization of the solution. This may inhibit some DNA polymerases, rendering subsequent PCR difficult. However, this situation can be avoided by monitoring the pH of the solution to ensure that desulphonation will be complete. A final concern is that bisulphite treatment greatly reduces the level of complexity in the sample, which can be problematic if multiple PCR reactions are to be performed (2006). Primer design is more difficult, and inappropriate cross-hybridization is more frequent.

Applications: Genome-wide Methylation Analysis

The advances in bisulfite sequencing have led to the possibility of applying them at a genome-wide scale, where, previously, global measure of DNA methylation was feasible only using other techniques, such as Restriction landmark genomic scanning. The mapping of the human epigenome is seen by many scientists as the logical follow-up to the completion of the Human Genome Project. This epigenomic information will be important in understanding how the function of the genetic sequence is implemented and regulated. Since the epigenome is less stable than the genome, it is thought to be important in gene-environment interactions.

Epigenomic mapping is inherently more complex than genome sequencing, however, since the epigenome is much more variable than the genome. While an individual has only one genome, one's epigenome varies with age, differs between tissues, is altered by environmental factors, and shows aberrations in diseases. Such rich epigenomic mapping, however, representing different ages, tissue types, and disease states, would yield valuable information on the normal function of epigenetic marks as well as the mechanisms leading to aging and disease.

Direct benefits of epigenomic mapping include probable advances in cloning technology. It is believed that failures to produce cloned animals with normal viability and lifespan result from inappropriate patterns of epigenetic marks.

Also, aberrant methylation patterns are well characterized in many cancers. Global hypomethylation results in decreased genomic stability, while local hypermethylation of tumour suppressor gene promoters often accounts for their loss of function. Specific patterns of methylation are indicative of specific cancer types, have prognostic value, and can help to guide the best course of treatment.

Large-scale epigenome mapping efforts are under way around the world and have been organized under the Human Epigenome Project. This is based on a multi-tiered strategy, whereby bisulfite sequencing is used to obtain high-resolution methylation profiles for a limited number of reference epigenomes, while less thorough analysis is performed on a wider spectrum of samples. This approach is intended to maximize the insight gained from a given amount of resources, as high-resolution genome-wide mapping remains a costly undertaking.

Expression Cloning

Expression cloning is a technique in DNA cloning that uses expression vectors to generate a library of clones, with each clone expressing one protein. This *expression library* is then screened for the property of interest and clones of interest recovered for further analysis. An example would be using an expression library to isolate genes that could confer antibiotic resistance.

Expression Vectors

Expression vectors are a specialized type of cloning vector in which the transcriptional and translational signals needed for the regulation of the gene of interest are included in the cloning vector. The transcriptional and translational signals may be synthetically created to make the expression of the gene of interest easier to regulate.

Purpose

Usually the ultimate aim of expression cloning is to produce large quantities of specific proteins. To this end, a bacterial expression clone may include a ribosome binding site (Shine-Dalgarno sequence) to enhance translation of the gene of interest's mRNA, a transcription termination sequence, or, in eukaryotes, specific sequences to promote the post-translational modification of the protein product.

Southern Blot

A Southern blot is a method routinely used in molecular biology for detection of a specific DNA sequence in DNA samples. Southern blotting combines transfer of electrophoresis-separated DNA fragments to a filter membrane and subsequent fragment detection by probe hybridization. The method is named after its inventor, the British biologist Edwin Southern.

Other blotting methods (i.e., Western blot, Northern blot, Eastern blot, Southwestern blot) that employ similar principles, but using RNA or protein, have later been named in reference to Edwin Southern's name. As the technique was eponymously named, Southern blot is capitalized as is conventional for proper nouns. The names for other blotting methods follow this convention by analogy.

Northern Blot

The northern blot is a technique used in molecular biology research to study gene expression by detection of RNA (or isolated mRNA) in a sample.

With northern blotting it is possible to observe cellular control over structure and function by determining the particular gene expression levels during differentiation, morphogenesis, as well as abnormal or diseased conditions. Northern blotting involves the use of electrophoresis to separate RNA samples by size, and detection with a hybridization probe complementary to part of or the entire target sequence. The term 'northern blot' actually refers specifically to the capillary transfer of RNA from the electrophoresis gel to the blotting membrane, however the entire process is commonly referred to as northern blotting. The northern blot technique was developed in 1977 by James Alwine, David Kemp, and George Stark at Stanford University. Northern blotting takes its name from its similarity to the first blotting technique, the Southern blot, named for biologist Edwin Southern. The major difference is that RNA, rather than DNA, is analysed in the Northern blot.

Procedure

A general blotting procedure starts with extraction of total RNA from a homogenized tissue sample. The mRNA can then be isolated through the use of oligo (dT) cellulose chromatography to maintain only those RNAs with a poly(A) tail. RNA samples are then separated by gel electrophoresis. Since the gels are fragile and the probes are unable to enter the matrix, the RNA samples, now separated by size, are transferred to a nylon membrane through a capillary or vacuum blotting system.

A nylon membrane with a positive charge is the most effective for use in northern blotting since the negatively charged nucleic acids have a high affinity for them. The transfer buffer used for the blotting usually contains formamide because it lowers the annealing temperature of the probe-RNA interaction preventing RNA degradation by high temperatures. Once the RNA has been transferred to the membrane it is immobilized through covalent linkage to the membrane by UV light or heat. After a probe has been labelled, it is hybridized to the RNA on the membrane. Experimental conditions that can affect the efficiency and specificity of hybridization include ionic strength, viscosity, duplex length, mismatched base pairs, and base composition. The membrane is washed to ensure that the probe has bound specifically and to avoid background signals from arising. The hybrid signals are then detected by X-ray film and can be quantified by densitometry. To create controls for comparison in a northern blot, samples not displaying the gene product of interest can be used after determination by microarrays or RT-PCR.

Gels

The RNA samples are most commonly separated on agarose gels containing formaldehyde as a denaturing agent for the RNA to limit secondary structure. The gels can be stained with ethidium bromide (EtBr) and viewed under UV light to observe the quality and quantity of RNA before blotting. Polyacrylamide gel electrophoeresis with urea can also be used in RNA separation but it is most commonly used for fragmented RNA or microRNAs. An RNA ladder is often run alongside the samples on an electrophoresis gel to observe the size of fragments obtained but in total RNA samples the ribosomal subunits can act as size markers. Since the large ribosomal subunit is 28S (approximately 5kb) and the small ribosomal subunit is 18S (approximately 2kb) two prominent bands will appear on the gel, the larger at close to twice the intensity of the smaller.

Probes

Probes for northern blotting are composed of nucleic acids with a complementary sequence to all or part of the RNA of interest, they can be DNA, RNA, or oligonucleotides with a minimum of 25 complementary bases to the target sequence. RNA probes (riboprobes) that are transcribed in vitro are able to withstand more rigorous washing steps preventing some of the background noise. Commonly cDNA is created with labelled primers for the RNA sequence of interest to act as the probe in the northern blot.

The probes need to be labelled either with radioactive isotopes or with chemiluminescence in which alkaline phosphatase or horseradish peroxidase breakdown chemiluminescent substrates producing a detectable emission of light. The chemiluminescent labelling can occur in two ways: either the probe is attached to the enzyme, or the probe is labelled with a ligand (e.g. biotin) for which the antibody (e.g. avidin or streptavidin) is attached to the enzyme. X-ray film can detect both the radioactive and chemiluminescent signals and many researchers prefer the chemiluminescent signals because they are faster, more sensitive, and reduce the health hazards that go along with radioactive labels. The same membrane can be probed up to five times without a significant loss of the target RNA.

Applications

Northern blotting allows one to observe a particular gene's expression pattern between tissues, organs, developmental stages, environmental stress levels, pathogen infection, and over the course of treatment. The technique has been used to show overexpression of oncogenes and downregulation of tumour-suppressor genes in cancerous cells when compared to 'normal' tissue, as well as the gene expression in the rejection of transplanted organs. If an upregulated gene is observed by an abundance of mRNA on the northern blot the sample can then be sequenced to determine if the gene is known to researchers or if it is a novel finding. The expression patterns obtained under given conditions can provide insight into the function of that gene. Since the RNA is first separated by size, if only one probe type is used variance in the level of each band on the membrane can provide insight into the size of the product, suggesting alternative splice products of the same gene or repetitive sequence motifs. The variance in size of a gene product can also indicate deletions or errors in transcript processing, by altering the probe target used along the known sequence it is possible to determine which region of the RNA is missing.

BlotBase is an online database publishing northern blots. BlotBase has over 700 published northern blots of human and mouse samples, in over 650 genes across more than 25 different tissue types. Northern blots can be searched by a blot ID, paper reference, gene identifier, or by tissue. The results of a search provide the blot ID, species, tissue, gene, expression level, blot image (if available), and links to the publication that the work originated from. This new database provides sharing of information between members of the science community that was not previously seen in northern blotting as it was in sequence analysis, genome determination, protein structure, etc.

Disadvantages and Advantages

Analysis of gene expression can be done by several different methods including RT-PCR, RNase protection assays, microarrays, serial analysis of gene expression (SAGE), as well as northern blotting. Microarrays are quite commonly used and are usually consistent with data obtained from northern blots; however, at times northern blotting is able to detect small changes in gene expression that microarrays cannot. The advantage that microarrays have over northern blots is that thousands of genes can be visualized at a time, while northern blotting is usually looking at one or a small number of genes.

A problem in northern blotting is often sample degradation by RNases (both endogenous to the sample and through environmental contamination), which can be avoided by proper sterilization of glassware and the use of RNase inhibitors such as DEPC (diethylpyrocarbonate). The chemicals used in most northern blots can be a risk to the researcher, since formaldehyde, radioactive material, ethidium bromide, DEPC, and UV light are all harmful under certain exposures. Compared to RT-PCR, northern blotting has a low sensitivity, but it also has a high specificity which is important to reduce false positive results.

The advantages of using northern blotting include the detection of RNA size, the observation of alternate splice products, the use of probes with partial homology, the quality and quantity of RNA can be measured on the gel prior to blotting, and the membranes can be stored and reprobed for years after blotting.

Reverse Northern Blot

A variant of the procedure known as the reverse northern blot is occasionally used. In this procedure, the substrate nucleic acid (that is affixed to the membrane) is a collection of isolated DNA fragments, and the probe is RNA extracted from a tissue and radioactively labelled.

The use of DNA microarrays that have come into widespread use in the late 1990s and early 2000s is more akin to the reverse procedure, in that they involve the use of isolated DNA fragments affixed to a substrate, and hybridization with a probe made from cellular RNA. Thus the reverse procedure, though originally uncommon, enabled northern analysis to evolve into gene expression profiling, in which many (possibly all) of the genes in an organism may have their expression monitored.

Sucrose Gradient Centrifugation

Sucrose gradient centrifugation is a type of centrifugation often used to purify enveloped viruses (with densities 1.1-1.2 g/cm^3) and ribosomes, and also to separate cell organelles from crude cellular extracts. This method is also used to purify exosomes.

Equilibrium Centrifugation

Typically, a sucrose density gradient is created by gently overlaying lower concentrations of sucrose on higher concentrations in a centrifuge tube. For example, a sucrose gradient may consist of layers extending from 70% sucrose to 20% sucrose in 10% increments (though this is highly variable depending on sample to be purified). The sample containing the particles of interest is placed on top of the gradient and centrifuged at forces in excess of 150,000 x g. The particles travel through the gradient until they reach the point in the gradient at which their density matches that of the surrounding sucrose. This fraction can then be removed and subjected to further analysis. A similar technique is sucrose cushion centrifugation, in which a particle mixture is pelleted through a 20% sucrose layer, coming to rest at the interface with a 70% solution. This allows concentration of particles from a sample. Unlike standard centrifugation, which in effect crushes the particles against the bottom of the centrifuge tube, the sucrose cushion method causes no mechanical stress and allows the collection of morphologically intact particles.

Non-equilibrium Centrifugation

This is very similar to the equilibrium form, but the experiment is only run until a particular point. Then the sucrose is eluted from the bottom of the tube.

Radioactivity in Biological Research

Radioactivity can be used in life sciences as a radiolabel to visualise components or target molecules in a biological system. Some radionuclei are synthesised in particle accelerators and have short half-lives, giving them high maximum theoretical specific activities. This lowers the detection time compared to radionuclei with longer half-lives, such as carbon-14. In some applications they have been substituted by fluorescent dyes.

Examples of Radionuclei

- Tritium (hydrogen-3) is a very low energy emitter that can be used to label proteins, nucleic acids, drugs and toxins, but requires a tritium-specific film or a tritium-specific phosphor screen. In a liquid scintillation assay (LSA), the efficiency is 20–50%, depending on the scintillation cocktail used. The maximum theoretical specific activity of tritium is 28.8 Ci/mmol (1.066 PBq/mol). However, there is often more than one tritium atom per molecule: for example, tritiated UTP is sold by most suppliers with carbons 5 and 6 each bonded to a tritium atom. C-14, S-35 and P-33 have similar emission energies. P-32 and I-125 are higher energy emitters -> inaccurate, see beta vs gamma radiation.
- Carbon-14 has a long half-life of 5,730±40 years. Its maximum specific activity is 0.0624 Ci/mmol (2.31 TBq/mol). It is used in applications such as radiometric dating or drug tests.
- Sodium-22 and chlorine-36 are commonly used to study ion transporters. However, sodium-22 is hard to screen off and chlorine-36, with a half-life of 300,000 years, has low activity.
- Sulfur-35 is used to label proteins and nucleic acids. Cysteine is an amino acid containing a thiol group which can be labelled by S-35. For nucleotides that do not contain a sulfur group, the oxygen on one of

the phosphate groups can be substituted with a sulfur. This thiophosphate acts the same as a normal phosphate group, although there is a slight bias against it by most polymerases. The maximum theoretical specific activity is 1,494 Ci/mmol (55.28 PBq/mol).

- Phosphorus-33 is used to label nucleotides. It is less energetic than P-32 and does not require protection with plexi glass. A disadvantage is its higher cost compared to P-32, as most of the bombarded P-31 will have acquired only one neutron, while only some will have acquired two or more. Its maximum specific activity is 5,118 Ci/mmol (189.4 PBq/mol).
- Phosphorus-32 is widely used for labelling nucleic acids and phosphoproteins. It has the highest emission energy (1.7 MeV) of all common research radioisotopes.

  This is a major advantage in experiments for which sensitivity is a primary consideration, such as titrations of very strong interactions (i.e., very low dissociation constant), footprinting experiments, and detection of low-abundance phosphorylated species. 32P is also relatively inexpensive. Because of its high energy, however, a number of safety and administrative controls are required (e.g., acrylic glass). The half-life of 32P is 14.2 days, and its maximum specific activity is 9131 Ci/mmol.
- Iodine-125 is commonly used for labelling proteins, usually at tyrosine residues. Unbound iodine is volatile and must be handled in a fume hood. Its maximum specific activity is 2,176 Ci/mmol (80.51 PBq/mol).

A good example of the difference in energy of the various radionuclei is the detection window rnages used to detect them, which are generally proportional to the energy of the emission, but vary from machine to machine: in a Perkin elmer TriLux Beta scintillation counter, the H-3 energy range window is between channel 5–360; C-14, S-35 and P-33 are in the window of 361–660; and P-32 is in the window of 661–1024.

Detection

Quantitative:

- In a liquid scintillation assay (LSA), or liquid scintillation counting (LSC), a small aliquot, filter or swab is added to scintillation fluid and the plate or vial counter in a scintillation counter.
- A Geiger counter is a quick and rough approximation of activity. Lower energy emitters such as tritium can not be detected.

Qualitative

- Autoradiography: A membrane such as a Northern blot or a hybridised slot blot is put against a film that is then developed.
- Phosphor storage screen: The membrane is placed against a phosphor storage screen which is then scanned in a phosphorimager. This is ten times faster and more precise than film and the result is already in digital form.

Microscopy

- Electron microscopy: The sample is not exposed to a beam of electrons but detectors picks up the expelled electrons from the radionuclei.
- Micro-autoradiography imager: A slide is put against scintillation paper and in a PMT. When two different radiolabels are used, a computer can be used to discriminate the two.

Scientific Methods

- Schild regression is a radioligand binding assay. It is used for DNA labelling (5' and 3'), leaving the nucleic acids intact.

Radioactivity Concentration

A vial of radiolabel has a "total activity". Taking as an example γ32P ATP, from the catalogues of the two major suppliers, Perkin Elmer NEG502H500UC or GE AA0068-500UCI, in this case, the total activity is 500 μCi (other typical numbers are 250 μCi or 1 mCi).

This is contained in a certain volume, depending on the radioactive concentration, such as 5 to 10 mCi/mL (185 to 370 TBq/m^3); typical volumes include 50 or 25 μL.

Not all molecules in the solution have a P-32 on the last (i.e., gamma) phosphate: the "specific activity" gives the radioactivity concentration and depends on the radionuclei's half-life. If every molecule were labelled, the maximum theoretical specific activity is obtained that for P-32 is 9131 Ci/mmol.

Due to pre-calibration and efficiency issues this number is never seen on a label; the values often found are 800, 3000 and 6000 Ci/mmol. With this number it is possible to calculate the total chemical concentration and the hot-to-cold ratio.

"Calibration date" is the date in which the vial's activity is the same as on the label. "Pre-calibration" is the when the activity is calibrated in a future date to compensate for the decay occurred during shipping.

Comparison with Fluorescence

Prior to the widespread use of fluorescence in the past three decades radioactivity was the most common label.

Advantages are:

- fluorescence is much safer and more convenient to use
- Several fluorescent molecules can be used simultaneously (given that they do not overlap, cf. FRET), whereas with radioactivity two isotopes can be used (tritium and a low energy isotope, e.g. ^{33}P due to different intensities) but require special machinery (a tritium screen and a regular phosphor-imaging screen or a specific dual channel detector, e.g.).
- Several properties are extremely useful (cf. next section)

Note: a channel is similar to "colour" but distinct, it is the pair of excitation and emission filters specific for a dye, e.g. agilent microarrays are dual channel, working on cy3 and cy5, these are colloquially referred to as green and red.

Disadvantages are:

- the dye may be a hindrance or toxic.

Safety

If good health physics controls are maintained in a laboratory where radionuclides are used, it is unlikely that the overall radiation dose received by workers will be of much significance. Nevertheless the effects of low doses are mostly unknown so many regulations exist to avoid unnecessary risks, such as skin or internal exposure. Due to the low penetration power and many variables involved it is hard to convert a radioactive concentration to a dose. 1 μCi of P-32 on a square centimetre of skin (through a dead layer of a thickness of 70 μm) gives 7961 rads (79.61 grays) per hour. Similarly a mammogram gives an exposure of 300 mrem (3 mSv) on a larger volume (in the US, the average annual dose is 360 mrem or 3.6 mSv).

Lab-on-a-chip

A lab-on-a-chip (LOC) is a device that integrates one or several laboratory functions on a single chip of only millimeters to a few square centimetres in size. LOCs deal with the handling of extremely small fluid volumes down to less than pico litres. Lab-on-a-chip devices are a subset of MEMS devices and often indicated by "Micro Total Analysis Systems" (μTAS) as well. Microfluidics is a broader term that describes also mechanical flow control devices like pumps and valves or sensors like flowmeters and viscometers. However, strictly regarded "Lab-on-a-Chip" indicates generally the scaling of single or multiple lab processes down to chip-format, whereas "μTAS" is dedicated to the integration of the total sequence of lab processes to perform chemical analysis. The term "Lab-on-a-Chip" was introduced later on when it turned out that μTAS technologies were more widely applicable than only for analysis purposes.

History

After the invention of microtechnology (~1954) for realizing integrated semiconductor structures for microelectronic chips, these lithography-based technologies were soon applied in pressure sensor manufacturing (1966) as well. Due to further development of these usually CMOS-compatibility limited processes, a tool box became available to create micrometre or sub-micrometre sized mechanical structures in silicon wafers as well: the Micro Electro Mechanical Systems (MEMS) era (also indicated with Micro System Technology - MST) had started.

Next to pressure sensors, airbag sensors and other mechanically movable structures, fluid handling devices were developed. Examples are: channels (capillary connections), mixers, valves, pumps and dosing devices. The first LOC analysis system was a gas chromatograph, developed in 1975 by S.C. Terry - Stanford University.

However, only at the end of the 1980s, and beginning of the 1990s, the LOC research started to seriously grow as a few research groups in Europe developed micropumps, flowsensors and the concepts for integrated fluid treatments for analysis systems. These μTAS concepts demonstrated that integration of pre-treatment steps, usually done at lab-scale, could extend the simple sensor functionality towards a complete laboratory analysis, including e.g. additional cleaning and separation steps.

A big boost in research and commercial interest came in the mid 1990's, when μTAS technologies turned out to provide interesting tooling for genomics applications, like capillary electrophoresis and DNA microarrays. A big boost in research support also came from the military, especially from DARPA (Defence Advanced Research Projects Agency), for their interest in portable bio/chemical warfare agent detection systems.

The added value was not only limited to integration of lab processes for analysis but also the characteristic possibilities of individual components and the application to other, non-analysis, lab processes. Hence the term "Lab-on-a-Chip" was introduced.

Although the application of LOCs is still novel and modest, a growing interest of companies and applied research groups is observed in different fields such as analysis (e.g. chemical analysis, environmental monitoring, medical diagnostics and cellomics) but also in synthetic chemistry (e.g. rapid screening and microreactors for pharmaceutics). Besides further application developments, research in LOC systems is expected to extend towards downscaling of fluid handling structures as well, by using nanotechnology. Sub-micrometre and nano-sized channels, DNA labyrinths, single cell detection and analysis, and nano-sensors, might become feasible, allowing new ways of interaction with biological species and large molecules.

Chip Materials and Fabrication Technologies

The basis for most LOC fabrication processes is photolithography. Initially most processes were in silicon, as these well-developed technologies were directly derived from semiconductor fabrication. Because of demands for e.g. specific optical characteristics, bio- or chemical compatibility, lower production costs and faster prototyping, new processes have been developed such as glass, ceramics and metal etching, deposition and bonding, PDMS processing (e.g., soft lithography), thick-film- and stereolithography as well as fast replication methods via electroplating, injection molding and embossing.

Furthermore the LOC field more and more exceeds the borders between lithography-based microsystem technology, nano technology and precision engineering.it is very useful.

Advantages of LOCs

LOCs may provide advantages, which are specific to their application. Typical advantages are:

- low fluid volumes consumption (less waste, lower reagents costs and less required sample volumes for diagnostics)
- faster analysis and response times due to short diffusion distances, fast heating, high surface to volume ratios, small heat capacities.
- better process control because of a faster response of the system (e.g. thermal control for exothermic chemical reactions)
- compactness of the systems due to integration of much functionality and small volumes
- massive parallelization due to compactness, which allows high-throughput analysis
- lower fabrication costs, allowing cost-effective disposable chips, fabricated in mass production
- safer platform for chemical, radioactive or biological studies because of integration of functionality, smaller fluid volumes and stored energies

Disadvantages of LOCs

- novel technology and therefore not yet fully developed
- physical and chemical effects—like capillary forces, surface roughness, chemical interactions of construction materials on reaction processes—become more dominant on small-scale. This can sometimes make processes in LOCs more complex than in conventional lab equipment
- detection principles may not always scale down in a positive way, leading to low signal-to-noise ratios
- although the absolute geometric accuracies and precision in microfabrication are high, they are often rather poor in a relative way, compared to precision engineering for instance.

Examples of LOC Applications

- Real-time PCR: detect bacteria, viruses and cancers.
- Biochemical assays
- Immunoassay: detect bacteria, viruses and cancers based on antigen-antibody reactions.
- Dielectrophoresis: detection of cancer cells and bacteria.
- Blood sample preparation: can crack cells to extract DNA.
- Cellular lab-on-a-chip for single-cell analysis.
- Ion channel screening (patch clamp)
- Testing the safety and efficacy of new drugs, as with lung on a chip

LOCs and Global Health

Lab-on-a-chip technology may soon become an important part of efforts to improve global health, particularly through the development of point-of-care testing devices. In countries with few healthcare resources, infectious diseases that would be treatable in a developed nation are often deadly. In some cases, poor healthcare clinics have the drugs to treat a certain illness but lack the diagnostic tools to identify patients who should receive the drugs. Many researchers believe that LOC technology may be the key to powerful new diagnostic instruments. The goal of these researchers is to create microfluidic chips that will allow healthcare providers in poorly equipped clinics to perform diagnostic tests such as immunoassays and nucleic acid assays with no laboratory support.

Global Challenges

For the chips to be used in areas with limited resources, many challenges must be overcome. In developed nations, the most highly valued traits for diagnostic tools include speed, sensitivity, and specificity; but in countries where the healthcare infrastructure is less well developed, attributes such ease of use and shelf life must also be considered. The reagents that come with the chip, for example, must be designed so that they remain effective for months even if the chip is not kept in a climate-controlled environment. Chip designers must also keep cost, scalability, and recyclability in mind as they choose what materials and fabrication techniques to use.

Examples of Global LOC Application

One active area of LOC research involves ways to diagnose and manage HIV infections. Around 40 million people are infected with HIV in the world today, yet only 1.3 million of these people receive anti-retroviral treatment. Around 90% of people with HIV have never been tested for the disease. Measuring the number of CD4+ T lymphocytes in a person's blood is an accurate way to determine if a person has HIV and to track the progress of an HIV infection. At the moment, flow cytometry is the gold standard for obtaining CD4 counts, but flow cytometry is a complicated technique that is not available in most developing areas because it requires trained technicians and expensive equipment.

Nuclear Run-on Assay

A nuclear run-on assay is conducted to identify the genes that are being transcribed at a certain time. Cell nuclei are isolated rapidly, and incubated with labelled nucleotides and the results are hybridized to a slot blot, which is then exposed to film. It was originally developed by Gariglio et al. (1981) and Brown et al. (1984). Adding actinomycin-D to the reaction buffer is sufficient to stop transcription, while other toxins, such as α-amanitin, show different effects, making it a good way to assay the effects of these compounds.

This method allows changes in transcription rates to be measured, which often differ from steady-state mRNA levels used in microarrays. Although the method is time-consuming and requires the use of radioactivity and large numbers of cells, it remains the most reliable method to measure transcription rates directly. Several attempts recently have been made to make this method microarray-compatible, but other methods are much more favourable. To have a publication quality blot about 10^7 nuclei are needed, but 10^6 nuclei are sufficient to give results: these quantities of cells are higher than for most protocols.

It is often cited that this method has been surpassed by microarrays, even though this method compares transcription rates and not total transcript expression levels, which differ. It is incompatible with microarray experiments due to the fact eukaryotic polymerases are more intolerant towards fluorescently labelled or aminoallyl nucleotides, so UTP with phosphorus-32, or phosphorus-

33, on the alpha phosphate group is normally used. Cheadle et al. (2005) used spotted nylon blots to obtain a high-throughput analysis of T-cell activation. Garcia-Martinez et al (2004)developed a protocol for the yeast S. cerevisiae (Genomic run-on, GRO) that allows for the calculation of transcription rates (TRs) for all yeast genes. Those authors also used the calculated TRs and the steady-state mRNA amounts of all genes to determine, at global scale, the mRNA stabilities for all yeast mRNAs. Alternative microarray methods have recently been developed, mainly PolII RIP-chip: RNA immunoprecipitation of RNA polymerase II with phosphorylated C-terminal domain directed antibodies and hybridization on a microarray slide or chip (the word chip in the name stems from "ChIP-chip" where a special affymetrix genechip was required). A comparison of methods based on run-on and ChIP-chip has been made in yeast (Pelechano et al., 2009).

A general correspondence of both methods has been detected but GRO is more sensitive and quantitative. It has to be considered that run-on only detects elongating RNA polymerases whereas ChIP-chip detects all present RNA polymerases, including backtracked ones. The main issue on the validity of run on assay results is the presence of artifacts due to the isolation step: some transcripts may stop prematurely others might start out of place. Another point to add is about the nuclei: There are various methods that require nuclear isolation. The problem is that the nuclear membrane is contiguous with the ER which is rich in protein and cytoplasmic mRNA; some methods require pure nuclear lysate while others, like a nuclear run-on, require "dirty" nuclei with an intact nuclear membrane even if the ER is attached. Some types of cell are notoriously difficult to isolate this way such as lymphocytes, which require a sucrose gradient purification. Yeast cells do not require nuclei isolation. Permeabilized cells (with sarkosil) can be used. This is a clear technical advantage.

Fluorescent in Situ Hybridization

FISH (fluorescence *in situ* hybridization) is a cytogenetic technique developed by Christoph Lengauer that is used to detect and localize the presence or absence of specific DNA sequences on chromosomes. FISH uses fluorescent probes that bind to only those parts of the chromosome with which they show a high degree of sequence similarity. Fluorescence microscopy can be used to find out where the fluorescent probe bound to the chromosomes. FISH is often used for finding specific features in DNA for use in genetic counselling, medicine, and species identification. FISH can also be used to detect and localize specific mRNAs within tissue samples. In this context, it can help define the spatial-temporal patterns of gene expression within cells and tissues.

Probes

Probes are often derived from fragments of DNA that were isolated, purified, and amplified for use in the Human Genome Project. The size of the human genome is so large, compared to the length that could be sequenced directly, that it was necessary to divide the genome into fragments. (In the eventual analysis, these fragments were put into order by digesting a copy of each fragment into still smaller fragments using sequence-specific endonucleases, measuring the size of each small fragment using size-exclusion chromatography, and using that information to determine where the large fragments overlapped one another.) To preserve the fragments with their individual DNA sequences, the fragments were added into a system of continually replicating bacteria populations. Clonal populations of bacteria, each population maintaining a single artificial chromosome, are stored in various laboratories around the world. The artificial chromosomes (BAC) can be grown, extracted, and labelled, in any lab. These fragments are on the order of 100 thousand base-pairs, and are the basis for most FISH probes.

Preparation and Hybridization Process

First, a probe is constructed. The probe must be large enough to hybridize specifically with its target but not so large as to impede the hybridization process. The probe is tagged directly with fluorophores, with targets for antibodies or with biotin. Tagging can be done in various ways, such as nick translation, or PCR using tagged nucleotides.

Then, an interphase or metaphase chromosome preparation is produced. The chromosomes are firmly attached to a substrate, usually glass. Repetitive DNA sequences must be blocked by adding short fragments of DNA to the sample. The probe is then applied to the chromosome DNA and incubated for approximately 12 hours while hybridizing. Several wash steps remove all unhybridized or partially-hybridized probes. The results are then visualized and quantified using a microscope that is capable of exciting the dye and recording images.

If the fluorescent signal is weak, amplification of the signal may be necessary in order to exceed the detection threshold of the microscope. Fluorescent signal strength depends on many factors such as probe labelling efficiency, the type of probe, and the type of dye. Fluorescently-tagged antibodies or streptavidin are bound to the dye molecule. These secondary components are selected so that they have a strong signal.

FISH experiments designed to detect or localize gene expression within cells and tissues rely on the use of a reporter gene, such as one expressing green fluorescent protein, to provide the fluorescence signal.

Variations on Probes and Analysis

FISH is a very general technique. The differences between the various FISH techniques are usually due to variations in the sequence and labelling of the probes; and how they are used in combination. These few modifications make possible all FISH techniques.

Probe size is important because longer probes hybridize more specifically than shorter probes. The overlap defines the resolution of detectable features. For example, if the goal of an experiment is to detect the breakpoint of a translocation, then the overlap of the probes — the degree to which one DNA sequence is contained in the adjacent probes — defines the minimum window in which the breakpoint may be detected.

The mixture of probe sequences determines the type of feature the probe can detect. Probes that hybridize along an entire chromosome are used to count the number of a certain chromosome, show translocations, or identify extra-chromosomal fragments of chromatin. This is often called "whole-chromosome painting." If every possible probe is used, every chromosome, (the whole genome) would be marked fluorescently, which would not be particularly useful for determining features of individual sequences. However, a mixture of smaller probes can be created that is specific to a particular region (locus) of DNA; these mixtures are used to detect deletion mutations. When combined with a specific colour, a locus-specific probe mixture is used to detect very specific translocations. Special locus-specific probe mixtures are often used to count chromosomes, by binding to the centromeric regions of chromosomes, which are unique enough to identify each chromosome (with the exception of Chromosome 13, 14 21, 22.)

A variety of other techniques use mixtures of differently-coloured probes. A range of colors in mixtures of fluorescent dyes can be detected, so each human chromosome can be identified by a characteristic colour using whole-chromosome probe mixtures and a variety of ratios of colors. Although there are more chromosomes than easily-distinguishable fluorescent dye colors, ratios of probe mixtures can be used to create *secondary* colors.

Similar to comparative genomic hybridization, the probe mixture for the secondary colours is created by mixing the correct ratio of two sets of differently-coloured probes for the same chromosome. This technique is sometimes called M-FISH. The same physics that make a variety of colours possible for M-FISH can be used for the detection of translocations. That is, colours that are adjacent appear to overlap; a secondary colour is observed. Some assays are designed so that the secondary colour will be present or absent in cases of interest. An example is the detection of BCR/ABL translocations, where the secondary colour indicates disease. This variation is often called double-fusion

FISH or D-FISH. In the opposite situation—where the absence of the secondary colour is pathological—is illustrated by an assay used to investigate translocations where only one of the breakpoints is known or constant. Locus-specific probes are made for one side of the breakpoint and the other intact chromosome. In normal cells, the secondary colour is observed, but only the primary colour is observed when the translocation occurs. This technique is sometimes called "break-apart FISH".

Single Molecule RNA FISH

Single Molecule RNA FISH is a method of detecting and quantifying mRNA and other long RNA molecules in a thin layer of tissue sample. Targets can be reliably imaged through the application of multiple short singly labelled oligonucleotide probes. The probes cooperatively bind to the target site. When each probe binds to the single stranded mRNA, it causes cooperative unwinding of the mRNA, promoting the binding of the next probe. The net result is the binding of 48 fluorescent labels to a single molecule of mRNA, providing sufficient fluorescence to reliably locate each target mRNA in a wide-field fluorescent microscopy image. Probes not binding to the intended sequence do not achieve sufficient localized fluorescence to be distinguished from the background. This technology is exclusively licensed to Biosearch Technologies as Stellaris™ FISH Probes. Single molecule RNA FISH assays can be performed in simplex or multiplex, and have potential applications in cancer diagnosis, neuroscience, gene expression analysis, and companion diagnostics.

Fiber FISH

In an alternative technique to interphase or metaphase preparations, fiber FISH, interphase chromosomes are attached to a slide in such a way that they are stretched out in a straight line, rather than being tightly coiled, as in conventional FISH, or adopting a random conformation, as in interphase FISH. This is accomplished by applying mechanical shear along the length of the slide, either to cells that have been fixed to the slide and then lysed, or to a solution of purified DNA. A technique known as chromosome combing is increasingly used for this purpose. The extended conformation of the chromosomes allows dramatically higher resolution - even down to a few kilobases. The preparation of fiber FISH samples, although conceptually simple, is a rather skilled art, and only specialized laboratories use the technique routinely.

Q-FISH

Q-FISH combines FISH with PNAs and computer software to quantify fluorescence intensity. This technique is used routinely in telomere length research.

Flow-FISH

Flow-FISH uses flow cytometry to perform FISH automatically using per-cell fluorescence measurements.

Medical Applications

Often parents of children with a developmental delay want to know more about their child's conditions before choosing to have another child. These concerns can be addressed by analysis of the parents' and child's DNA. In cases where the child's developmental delay is not understood, the cause of it can be determined using FISH and cytogenetic techniques.

Examples of diseases that are diagnosed using FISH include Prader-Willi syndrome, Angelman syndrome, 22q13 deletion syndrome, chronic myelogenous leukemia, acute lymphoblastic leukemia, Cri-du-chat, Velocardiofacial syndrome, and Down syndrome. FISH on sperm cells is indicated for men with an abnormal somatic or meiotic karyotype as well as those with oligozoospermia, since approximately 50% of oligozoospermic men have an increased rate of sperm chromosome abnormalities. The analysis of chromosomes 21, X, and Y is enough to identify oligozoospermic individuals at risk.

In medicine, FISH can be used to form a diagnosis, to evaluate prognosis, or to evaluate remission of a disease, such as cancer. Treatment can then be specifically tailored. A traditional exam involving metaphase chromosome analysis is often unable to identify features that distinguish one disease from another, due to subtle chromosomal features; FISH can elucidate these differences.

FISH can also be used to detect diseased cells more easily than standard Cytogenetic methods, which require dividing cells and requires labour and time-intensive manual preparation and analysis of the slides by a technologist. FISH, on the other hand, does not require living cells and can be quantified automatically, a computer counts the fluorescent dots present. However, a trained technologist is required to distinguish subtle differences in banding patterns on bent and twisted metaphase chromosomes.

Species Identification

FISH is often used in clinical studies. If a patient is infected with a suspected pathogen, bacteria, from the patient's tissues or fluids, are typically grown on agar to determine the identity of the pathogen. Many bacteria, however, even well-known species, do not grow well under laboratory conditions. FISH can be used to detect directly the presence of the suspect on small samples of patient's tissue.

FISH can also be used to compare the genomes of two biological species, to deduce evolutionary relationships. A similar hybridization technique is called a zoo blot. Bacterial FISH probes are often primers for the 16s rRNA region.

FISH is widely used in the field of microbial ecology, to identify microorganisms. Biofilms, for example, are composed of complex (often) multi-species bacterial organizations. Preparing DNA probes for one species and performing FISH with this probe allows one to visualize the distribution of this specific species within the biofilm. Preparing probes (in two different colours) for two species allows to visualize/study co-localization of these two species in the biofilm, and can be useful in determining the fine architecture of the biofilm.

Lab-on-a-chip and FISH

Although interphase fluorescence in situ hybridization (FISH) is a sensitive diagnostic tool used for the detection of chromosomal abnormalities on cell-by-cell basis, the cost-per-test and the technical complexity of current FISH protocols has inhibited its widespread utilization. Lab-on-a-chip or microfluidic devices, incorporate networks of microchannels that can miniaturize, integrate and automate conventional analytical techniques onto chip-style platforms. Since microchannels permit sophisticated levels of fluid control (down to picolitres), these devices can reduce analysis times, lower reagent consumption, and minimize human intervention.

Currently, FISH has been performed on glass microfluidic platforms that standardize much of the protocol offering repeatable results that are accurate, cost-effective and easier to obtain in a clinical setting. Compared to conventional FISH methods, these first implementations of on-chip FISH provide a 10-fold higher throughput and a 10-fold reduction in the cost of testing, enabling the simultaneous assessment of several chromosomal abnormalities or patients. It is increasingly essential that diagnostic tests determine the type and extent of chromosomal abnormalities for more informed diagnosis and for appropriate choice of treatment strategies. Since the on-chip FISH technique is 10-20 times more cost-effective than conventional methods, and can be fully integrated and automated, this technology will make widespread genetic testing of patients more accessible in a clinical setting.

Virtual Karyotype

Virtual karyotyping is another cost-effective, clinically available alternative to FISH panels uses thousands to millions of probes on a single array to detect copy number changes, genome-wide, at unprecedented resolution.

Bioinformatics

Bioinformatics is the application of statistics and computer science to the field of molecular biology.

The term *bioinformatics* was coined by Paulien Hogeweg and Ben Hesper in 1978 for the study of informatic processes in biotic systems. Its primary use since at least the late 1980s has been in genomics and genetics, particularly in those areas of genomics involving large-scale DNA sequencing.

Bioinformatics now entails the creation and advancement of databases, algorithms, computational and statistical techniques and theory to solve formal and practical problems arising from the management and analysis of biological data. Over the past few decades rapid developments in genomic and other molecular research technologies and developments in information technologies have combined to produce a tremendous amount of information related to molecular biology. It is the name given to these mathematical and computing approaches used to glean understanding of biological processes.

Common activities in bioinformatics include mapping and analysing DNA and protein sequences, aligning different DNA and protein sequences to compare them and creating and viewing 3-D models of protein structures. The primary goal of bioinformatics is to increase the understanding of biological processes. What sets it apart from other approaches, however, is its focus on developing and applying computationally intensive techniques (e.g., pattern recognition, data mining, machine learning algorithms, and visualization) to achieve this goal. Major research efforts in the field include sequence alignment, gene finding, genome assembly, drug design, drug discovery, protein structure alignment, protein structure prediction, prediction of gene expression and protein-protein interactions, genome-wide association studies and the modeling of evolution.

Introduction

Bioinformatics was applied in the creation and maintenance of a database to store biological information at the beginning of the "genomic revolution", such as nucleotide and amino acid sequences. Development of this type of database involved not only design issues but the development of complex interfaces whereby researchers could both access existing data as well as submit new or revised data.

In order to study how normal cellular activities are altered in different disease states, the biological data must be combined to form a comprehensive picture of these activities. Therefore, the field of bioinformatics has evolved such that the most pressing task now involves the analysis and interpretation of various types of data, including nucleotide and amino acid sequences, protein domains, and protein structures.

The actual process of analysing and interpreting data is referred to as computational biology. Important sub-disciplines within bioinformatics and computational biology include:

- the development and implementation of tools that enable efficient access to, and use and management of, various types of information.
- the development of new algorithms (mathematical formulas) and statistics with which to assess relationships among members of large data sets, such as methods to locate a gene within a sequence, predict protein structure and/or function, and cluster protein sequences into families of related sequences.

There are two fundamental ways of modelling a Biological system (e.g. living cell) both coming under Bioinformatic approaches.

- Static
  - — Sequences - Proteins, Nucleic acids and Peptides
  - — Structures - Proteins, Nucleic acids, Ligands (including metabolites and drugs) and Peptides
  - — Interaction data among the above entities including microarray data and Networks of proteins, metabolites
- Dynamic
  - o Systems Biology comes under this category including reaction fluxes and variable concentrations of metabolites
  - o Multi-Agent Based modelling approaches capturing cellular events such as signalling, transcription and reaction dynamics.

A broad sub-category under bioinformatics is structural bioinformatics.

Major Research Areas

Sequence Analysis

Since the Phage Φ-X174 was sequenced in 1977, the DNA sequences of thousands of organisms have been decoded and stored in databases. This sequence information is analysed to determine genes that encode polypeptides (proteins), RNA genes, regulatory sequences, structural motifs, and repetitive sequences. A comparison of genes within a species or between different species can show similarities between protein functions, or relations between species (the use of molecular systematics to construct phylogenetic trees). With the growing amount of data, it long ago became impractical to analyze DNA sequences manually. Today, computer programs such as BLAST are used daily to search sequences from more than 260 000 of organisms, containing over 190 billion nucleotides. These programs can compensate for mutations (exchanged, deleted or inserted bases) in the DNA sequence, in order to identify sequences that are related, but not identical. A variant of this sequence alignment is used in the sequencing process itself. The so-called shotgun sequencing technique (which was used, for example, by The Institute for Genomic Research to sequence the first bacterial genome, *Haemophilus influenzae*) does not produce entire chromosomes, but instead generates the sequences of many thousands of small DNA fragments (ranging from 35 to 900 nucleotides long, depending on the sequencing technology). The ends of these fragments overlap and, when aligned properly by a genome assembly program, can be used to reconstruct the complete genome. Shotgun sequencing yields sequence data quickly, but the task of assembling the fragments can be quite complicated for larger genomes. For a genome as large as the human genome, it may take many days of CPU time on large-memory, multiprocessor computers to assemble the fragments, and the resulting assembly will usually contain numerous gaps that have to be filled in later. Shotgun sequencing is the method of choice for virtually all genomes sequenced today, and genome assembly algorithms are a critical area of bioinformatics research.

Another aspect of bioinformatics in sequence analysis is annotation, which involves computational gene finding to search for protein-coding genes, RNA genes, and other functional sequences within a genome. Not all of the nucleotides within a genome are part of genes. Within the genome of higher organisms, large parts of the DNA do not serve any obvious purpose. This so-called junk DNA may, however, contain unrecognized functional elements. Bioinformatics helps to bridge the gap between genome and proteome projects — for example, in the use of DNA sequences for protein identification.

Genome Annotation

In the context of genomics, annotation is the process of marking the genes and other biological features in a DNA sequence. The first genome annotation software system was designed in 1995 by Dr. Owen White, who was part of the team at The Institute for Genomic Research that sequenced and analysed the first genome of a free-living organism to be decoded, the bacterium *Haemophilus influenzae*. Dr. White built a software system to find the genes (places in the DNA sequence that encode a protein), the transfer RNA, and other features, and to make initial assignments of function to those genes. Most current genome annotation systems work similarly, but the programs available for analysis of genomic DNA are constantly changing and improving.

Computational Evolutionary Biology

Evolutionary biology is the study of the origin and descent of species, as well as their change over time. Informatics has assisted evolutionary biologists in several key ways; it has enabled researchers to:

- trace the evolution of a large number of organisms by measuring changes in their DNA, rather than through physical taxonomy or physiological observations alone,
- more recently, compare entire genomes, which permits the study of more complex evolutionary events, such as gene duplication, horizontal gene transfer, and the prediction of factors important in bacterial speciation,
- build complex computational models of populations to predict the outcome of the system over time
- track and share information on an increasingly large number of species and organisms

Future work endeavours to reconstruct the now more complex tree of life.

The area of research within computer science that uses genetic algorithms is sometimes confused with computational evolutionary biology, but the two areas are not necessarily related.

Analysis of Gene Expression

The expression of many genes can be determined by measuring mRNA levels with multiple techniques including microarrays, expressed cDNA sequence tag (EST) sequencing, serial analysis of gene expression (SAGE) tag sequencing, massively parallel signature sequencing (MPSS), or various applications of multiplexed in-situ hybridization. All of these techniques are extremely noise-prone and/or subject to bias in the biological measurement, and a major research area in computational biology involves developing statistical tools to separate signal from noise in high-throughput gene expression studies. Such studies are often used to determine the genes implicated in a disorder: one might compare microarray data from cancerous epithelial cells to data from non-cancerous cells to determine the transcripts that are up-regulated and down-regulated in a particular population of cancer cells.

Analysis of Regulation

Regulation is the complex orchestration of events starting with an extracellular signal such as a hormone and leading to an increase or decrease in the activity of one or more proteins. Bioinformatics techniques have been applied to explore various steps in this process. For example, promoter analysis involves the identification and study of sequence motifs in the DNA surrounding the coding region of a gene. These motifs influence the extent to which that region is transcribed into mRNA. Expression data can be used to infer gene regulation: one might compare microarray data from a wide variety of states of an organism to form hypotheses about the genes involved in each state. In a single-cell organism, one might compare stages of the cell cycle, along with various stress conditions (heat shock, starvation, etc.). One can then apply clustering algorithms to that expression data to determine which genes are co-expressed. For example, the upstream regions (promoters) of co-expressed genes can be searched for over-represented regulatory elements.

Analysis of Protein Expression

Protein microarrays and high throughput (HT) mass spectrometry (MS) can provide a snapshot of the proteins present in a biological sample. Bioinformatics is very much involved in making sense of protein microarray and HT MS data; the former approach faces similar problems as with microarrays targeted at mRNA, the latter involves the problem of matching large amounts of mass data against predicted masses from protein sequence databases, and the complicated statistical analysis of samples where multiple, but incomplete peptides from each protein are detected.

Analysis of Mutations in Cancer

In cancer, the genomes of affected cells are rearranged in complex or even unpredictable ways. Massive sequencing efforts are used to identify previously unknown point mutations in a variety of genes in cancer. Bioinformaticians continue to produce specialized automated systems to manage the sheer volume of sequence data produced, and they create new algorithms and software to compare the sequencing results to the growing collection of human genome sequences and germline polymorphisms. New physical detection technologies are employed, such as oligonucleotide microarrays to identify chromosomal gains and losses (called comparative genomic hybridization), and single-nucleotide polymorphism arrays to detect known *point mutations*. These detection methods simultaneously measure several hundred thousand sites throughout the genome, and when used in high-throughput to measure thousands of samples, generate terabytes of data per experiment. Again the massive amounts and new types of data generate new opportunities for bioinformaticians. The data is often found to contain considerable variability, or noise, and thus Hidden Markov model and change-point analysis methods are being developed to infer real copy number changes.

Another type of data that requires novel informatics development is the analysis of lesions found to be recurrent among many tumours.

Comparative Genomics

The core of comparative genome analysis is the establishment of the correspondence between genes (orthology analysis) or other genomic features in different organisms. It is these intergenomic maps that make it possible to trace the evolutionary processes responsible for the divergence of two genomes. A multitude of evolutionary events acting at various organizational levels shape genome evolution. At the lowest level, point mutations affect individual nucleotides. At a higher level, large chromosomal segments undergo duplication, lateral transfer, inversion, transposition, deletion and insertion. Ultimately, whole genomes are involved in processes of hybridization, polyploidization and endosymbiosis, often leading to rapid speciation. The complexity of genome evolution poses many exciting challenges to developers of mathematical models and algorithms, who have recourse to a spectra of algorithmic, statistical and mathematical techniques, ranging from exact, heuristics, fixed parameter and approximation algorithms for problems based on parsimony models to Markov Chain Monte Carlo algorithms for Bayesian analysis of problems based on probabilistic models.

Many of these studies are based on the homology detection and protein families computation.

Modeling Biological Systems

Systems biology involves the use of computer simulations of cellular subsystems (such as the networks of metabolites and enzymes which comprise metabolism, signal transduction pathways and gene regulatory networks) to both analyze and visualize the complex connections of these cellular processes. Artificial life or virtual evolution attempts to understand evolutionary processes via the computer simulation of simple (artificial) life forms.

High-throughput Image Analysis

Computational technologies are used to accelerate or fully automate the processing, quantification and analysis of large amounts of high-information-content biomedical imagery. Modern image analysis systems augment an observer's ability to make measurements from a large or complex set of images, by improving accuracy, objectivity, or speed. A fully developed analysis system may completely replace the observer. Although these systems are not unique to biomedical imagery, biomedical imaging is becoming more important for both diagnostics and research. Some examples are:

- high-throughput and high-fidelity quantification and sub-cellular localization (high-content screening, cytohistopathology, Bioimage informatics)
- morphometrics
- clinical image analysis and visualization
- determining the real-time air-flow patterns in breathing lungs of living animals

- quantifying occlusion size in real-time imagery from the development of and recovery during arterial injury
- making behavioural observations from extended video recordings of laboratory animals
- infrared measurements for metabolic activity determination
- inferring clone overlaps in DNA mapping, e.g. the Sulston score.

Structural Bioinformatic Approaches

Prediction of Protein Structure

Protein structure prediction is another important application of bioinformatics. The amino acid sequence of a protein, the so-called primary structure, can be easily determined from the sequence on the gene that codes for it. In the vast majority of cases, this primary structure uniquely determines a structure in its native environment. (Of course, there are exceptions, such as the bovine spongiform encephalopathy - aka Mad Cow Disease - prion.) Knowledge of this structure is vital in understanding the function of the protein. For lack of better terms, structural information is usually classified as one of *secondary*, *tertiary* and *quaternary* structure. A viable general solution to such predictions remains an open problem. As of now, most efforts have been directed towards heuristics that work most of the time.

One of the key ideas in bioinformatics is the notion of homology. In the genomic branch of bioinformatics, homology is used to predict the function of a gene: if the sequence of gene *A*, whose function is known, is homologous to the sequence of gene *B,* whose function is unknown, one could infer that B may share A's function. In the structural branch of bioinformatics, homology is used to determine which parts of a protein are important in structure formation and interaction with other proteins.

In a technique called homology modeling, this information is used to predict the structure of a protein once the structure of a homologous protein is known. This currently remains the only way to predict protein structures reliably.

One example of this is the similar protein homology between hemoglobin in humans and the hemoglobin in legumes (leghemoglobin). Both serve the same purpose of transporting oxygen in the organism. Though both of these proteins have completely different amino acid sequences, their protein structures are virtually identical, which reflects their near identical purposes.

Other techniques for predicting protein structure include protein threading and *de novo* (from scratch) physics-based modeling.

Molecular Interaction

Efficient software is available today for studying interactions among proteins, ligands and peptides. Types of interactions most often encountered in the field include - Protein-ligand (including drug), protein-protein and protein-peptide.

Molecular dynamic simulation of movement of atoms about rotatable bonds is the fundamental principle behind computational algorithms, termed docking algorithms for studying molecular interactions.

Docking Algorithms

In the last two decades, tens of thousands of protein three-dimensional structures have been determined by X-ray crystallography and Protein nuclear magnetic resonance spectroscopy (protein NMR). One central question for the biological scientist is whether it is practical to predict possible protein-protein interactions only based on these 3D shapes, without doing protein-protein interaction experiments. A variety of methods have been developed to tackle the Protein-protein docking problem, though it seems that there is still much work to be done in this field.

Software and Tools

Software tools for bioinformatics range from simple command-line tools, to more complex graphical programs and standalone web-services available from various bioinformatics companies or public institutions.

Web Services in Bioinformatics

SOAP and REST-based interfaces have been developed for a wide variety of bioinformatics applications allowing an application running on one computer in one part of the world to use algorithms, data and computing resources on servers in other parts of the world. The main advantages derive from the fact that end users do not have to deal with software and database maintenance overheads. Basic bioinformatics services are classified by the EBI into three categories: SSS (Sequence Search Services), MSA (Multiple Sequence Alignment) and BSA (Biological Sequence Analysis).

The availability of these service-oriented bioinformatics resources demonstrate the applicability of web based bioinformatics solutions, and range from a collection of standalone tools with a common data format under a single, standalone or web-based interface, to integrative, distributed and extensible bioinformatics workflow management systems.

2

Enzymes in Molecular Biology Research

Since molecular cloning has become routine laboratory technique, manufacturers offer countless sources of enzymes to generate and manipulate nucleic acids. Thus, selecting the appropriate enzyme for a specific task may seem difficult to the novice. This review aims at providing the readers with some cues for understanding the function and specificities of the different sources of polymerases, ligases, nucleases, phosphatases, methylases, and topoisomerases used for molecular cloning. We provide a description of the most commonly used enzymes of each group, and explain their properties and mechanism of action. By pointing out key requirements for each enzymatic activity and clarifying their limitations, we aim at guiding the reader in selecting appropriate enzymatic source and optimal experimental conditions for molecular cloning experiments.

Keywords: Enzymes, Molecular biology, Molecular cloning.

At a time when molecular cloning has become routine laboratory technique, we thought it was important to provide readers with some cues for understanding the function and specificities of the different enzymes used to generate and manipulate nucleic acids. Over the past few years, the tremendous expansion of cloning techniques and applications has triggered an enormous interest from laboratory suppliers. As a result, countless sources of enzymes are now available, and selecting the appropriate enzyme for a specific task may seem difficult to the novice.

Nucleic acids used for molecular cloning can be of natural or synthetic origin, and their length ranges from a few to several thousands nucleotides.

Nucleic acids can be extensively manipulated, in order to acquire specific characteristics and properties. Such manipulations include propagation, ligation, digestion, or addition of modifying groups such as phosphate or methyl groups. These modifications are catalyzed by polymerases, ligases, nucleases, phosphatases, and methylases, respectively. In this review, we provide a description of the main enzymes of each group, and explain their properties and mechanism of action. Our goal is to give the reader a better understanding of the fundamental enzymatic activities that are used in molecular cloning.

DNA Polymerases (DNA-dependent DNA Polymerase, EC 2.7.7.7)

DNA polymerases are enzymes that catalyze the formation of polymers made by the assembly of multiple structural units or deoxyribo-nucleotides triphosphate (dNTPs). None of the DNA polymerases that have been characterized thus far can direct *de novo* synthesis of a polynucleotidic molecule from individual nucleotides. The DNA polymerases only add nucleotides to the 32 -OH end of a pre-existing primer containing a 52 -phosphate group. Primers are short stretches of RNA complementary to about 10 nucleotides of DNA at the 52 end of the molecule to replicate. Primers are synthesized by an RNA polymerase called primase.

Most polymerases used in molecular biology originate from bacteria or their infecting viruses (bacteriophages or phages). We will only discuss prokaryotic polymerases in this review. The functional and structural properties of eukaryotic DNA polymerases, which are specific for chromatin-embedded DNA, are reviewed elsewhere (Frouin et al. 2003; Garg and Burgers 2005).

Prokaryotic DNA Polymerases

In bacteria, three DNA polymerases act in concert to achieve DNA replication: Pol I, Pol II, and Pol III. All three enzymes catalyze 52 '!32 elongation of DNA strands in the presence of primer and dNTPs, but have variable elongation rate. DNA Pol III was discovered in 1970 (Kornberg and Gefter 1970) and is the main enzymatic complex driving prokaryotic replication. All three DNA Pol also have a 32 $\rightarrow$ 52 (reverse) exonuclease activity, otherwise known as proofreading activity as it initiates removal of incorrectly added bases as polymerization progresses. The proofreading activity increases fidelity but slows down polymerase progression. In addition to polymerase and proofreading activities common to all three DNA polymerases, DNA Pol I also has a 52 '!32 exonuclease activity, which is used for the removal of RNA primer from the 52 end of DNA chains, and for excision-repair (upstream of polymerization). The 52 '!32 exonuclease activity of Pol I is utilized *in vitro* for nick-translation (*i.e.* tagging technique in which some of the nucleotides of a DNA sequence are replaced with labeled analogues).

Pol I and Klenow Fragment

The native DNA Pol I has been successfully used to remove 32 protruding DNA ends (in the absence of dNTPs), or to fill in cohesive ends (in the presence of dNTPs) before addition of molecular linkers. However, the 52 '!32 exonuclease activity of DNA Pol I makes it unsuitable for all applications that require polymerization activity alone (*e.g.* to fill in cohesive ends before addition of linkers, or to copy single-stranded DNA in the dideoxy method for sequencing). Fortunately, it was discovered that proteolytic digestion of *E. Coli* DNA Pol I (109 kDa) generates two fragments (76 and 36 kDa). The large fragment, also known as DNA Pol IK or Klenow fragment (named after its inventor, (Klenow and Henningsen 1970)), contains the 52 '!32 polymerase and 32 → 52 exonuclease (proofreading) activities of DNA Pol I, while the small fragment exhibits the 52 '!32 exonuclease activity alone. Since then, recombinant sources of Klenow significantly improved the functional quality of this fragment by eliminating contaminations due to the presence of residual native enzyme in proteolytically treated preparations.

T4 DNA Polymerase

Bacteriophage T4 polymerase requires a template and a primer to exhibits two activities: it is a 32 '!52 (reverse) exonuclease in the absence of dNTPs, and a 52 '!32 polymerase in the presence of dNTPs. Unlike the *E. Coli* DNA Pol I, T4 DNA Pol does not exhibit a 52 '!32 exonuclease activity (Englund 1971). Therefore, T4 DNA Pol can be used instead of Klenow to fill in 52 -protruding ends of DNA fragments, for nick translation, or for labelling 32 ends of duplex (double-stranded) DNA. T4 DNA Pol exonuclease rate is approximately 40 bases per minutes on double stranded DNA, and about 4,000 bases on single stranded DNA. T4 DNA Pol polymerization rate reaches 15,000 nucleotides per minutes when assayed under standardized conditions.

Modified Bacteriophage T7 DNA Polymerase

A chemically modified phage T7 DNA polymerase has been described by Richardson et al. as an ideal tool for DNA sequencing (Huber et al. 1987; Tabor et al. 1987; Tabor and Richardson 1987). Modified bacteriophage T7 DNA Pol is a complex of two proteins: the 84 kDa product of the T7 gene 5, and the 12 kDa *E. Coli* thioredoxin (Mark and Richardson 1976; Modrich and Richardson 1975).

The T7 gene 5 protein provides catalytic properties to the complex, while the thioredoxin protein connects the T7 gene 5 protein to a primer template, which allows the polymerization of thousands of nucleotides without dissociation, thereby increasing the efficiency of the T7 polymerase. Hence, modified T7 DNA Pol has a polymerization rate of more than 300 nucleotides per second, which makes it more than 70 times faster than that of AMV reverse transcriptase. Thus, this enzymatic complex can be used for preparation of radioactive probes and amplification of large DNA fragments.

Further characterization identified a 28-amino acid region (residues 118–145) as essential to T7 DNA Pol 32 '!52 exonuclease activity (Tabor and Richardson 1989). *In vitro* mutagenesis of the corresponding nucleotides in the T7 gene 5 resulted in complete elimination of the exonuclease activity, thereby increasing polymerase efficiency (9-fold) and spontaneous mutation rate (14-fold) (Tabor and Richardson 1989). The mutant T7 polymerase/thioredoxin complex, commercially available under the name sequenase (United States Biochemical Corporation), is used for DNA sequencing because of its high efficiency and ability to incorporate nucleotide analogs (such as 52 -(α-thio)-dNTPs, dc7-GTP, or dITP used to improve the resolution of DNA sequencing gels, and to avoid gel compression resulting from base pairing).

Terminal Deoxynucleotidyl Transferase (DNA Nucleotidyl-exotransferase, EC 2.7.7.31)

Terminal deoxynucleotidyl transferase (TdT), initially purified from calf thymus (Krakow et al. 1962), is a DNA polymerase that catalyzes the addition of a homopolymer tail to 32 -OH ends of DNA, in a template-independent manner. TdT is used in molecular biology for labelling DNA 32 ends with modified nucleotides (such as ddNTP, DIG-dUTP, or radiolabeled nucleotides), for primer extension, or for DNA sequencing. It is also used in TUNEL (TdT dUTP Nick End Labelling) assay for the demonstration of apoptosis (Gavrieli et al. 1992).

TdT requires a single-stranded DNA primer in the presence of Mg^{2+} (three nucleotide-long minimum), but can accept double-stranded DNA as a primer in the presence of cobalt ions (Roychoudhury et al. 1976). However, addition of Co^{2+} may result in a relaxation of the helical structure of the DNA, thereby allowing the tailing of internal nicks. Use of Mg^{2+} reduces this problem but results in a significant decrease of the tail length (approximately one-fifth of the length obtained in the presence of Co^{2+} at identical enzyme:DNA ratios). Hence, optimization of incubation conditions is critical to control specificity, reactivity, and activity of TdT.

Thermostable DNA Polymerases

Although thermostable DNA polymerases were purified in the early seventies, their considerable interest for molecular cloning emerged from the development of polymerase chain reaction (PCR) and subsequent need for enzymes able to perform DNA synthesis at high temperature. Since thermostable polymerases are functional at high temperature, they can replicate DNA regions with high G/C content, similar to those frequently found in thermophilic organisms (high C/G content sequences form secondary structures that need to be properly denatured in order to be efficiently copied).

Bst Polymerase

Bst polymerase is a thermostable DNA Pol I that was isolated in 1968 (Stenesh et al. 1968) from the thermophilic bacterium *Bacillus stearothermophilus* (Bst), which proliferates between 39 and 70°C. Bst Pol I is active at an optimal temperature of 65°C, and is inactivated after 15 min incubation at 75°C. Bst Pol I possesses a 52 '!32 exonuclease activity and requires high Mg^{2+} concentration for maximum activity.

Protease digestion of Bst DNA Pol I generates two protein fragments. The large protein fragment of Bst DNA Pol I is thermostable, and thus very useful for sequencing reactions performed at 65°C to avoid problems due to hairpin formation. Like Klenow, the Large Fragment of Bst DNA Pol I shows a faster strand displacement than its full length counterpart. Recombinant Bst DNA Pol I is presently available from Epicentre Biotechnologies, which also commercializes the Large Fragment of rBst DNA Pol I under the name IsoTherm™.

Taq Polymerase

Taq polymerase was first purified in 1976 (Chien et al. 1976) from a bacterium discovered 8 years earlier in the Great Fountain region of Yellowstone National Park. This bacterium, which thrives at 70°C and survives at temperatures as high as 80°C, was named *Thermophilus aquaticus* (*T. aquaticus* or Taq). Taq polymerase has a halflife of 40 min at 95°C, and 5 min at 100°C, which allows the PCR reaction. For optimal activity, Taq polymerase requires Mg^{2+} and a temperature of 80°C. Taq polymerase lacks 32 '!52 exonuclease (proofreading) activity, and is therefore often described as being a low fidelity enzyme (error rate between 1×10^{-4} and 2×10^{-5} errors per base pair, depending on experimental conditions). Yet, one should keep in mind that this corresponds to a quite good accuracy (inverse of the error rate) since 45,000 nucleotides can be incorporated into newly synthesized DNA strands before an error occurs.

Like other DNA polymerases lacking 32 '!52 exonuclease activity, Taq polymerase exhibits a deoxynucleotidyl transferase activity that is accountable for the addition of a few adenine residues at the 32 -end of PCR products. Later on, additional thermostable DNA polymerases have been discovered and commercialized. These enzymes have better accuracy than the original Taq. However, the term "Taq" is commonly used in place of "thermostable DNA polymerase", hence the erroneous term "high fidelity Taq", often used in laboratories.

Tth Polymerase

Tth polymerase, isolated from *Thermus thermophilus HB-8* (Ruttimann et al. 1985), is 94 kDa thermostable polymerase lacking 32 '!52 exonuclease (proofreading) activity. Tth polymerase catalyzes the polymerization of nucleotides into duplex DNA from a DNA template (DNA polymerase) in the presence of Mg^{2+}, and from an RNA template (reverse transcriptase) in the presence of Mn^{2+}.

Thermostable DNA Polymerases with Proofreading Activity

Several DNA polymerases isolated from thermophillic organisms exhibit 32 '!52 exonuclease (proofreading) activity, which increases fidelity. Among them, DNA polymerase from the archebacteirum *Pyrococcus furiosus* (Pfu) has an error rate ($1.6\times10^{"6}$) 10-fold lower than that of Taq polymerase. It can be purchased from numerous providers. Pow polymerase (isolated from *Pyrococcus woesei*) has a half-life greater than 2 h at 100°C and a very low error rate of 7.4×10^{-7}. Pow polymerase provides high PCR yields only for templates shorter than 3.5 kb. To circumvent PCR product size limitation, a mixture of Taq and Pow DNA polymerases has been commercialized by Roche Molecular Chemicals (under the name "Expand high fidelity"). This system allows amplification and cloning of long stretches of DNA (20–35 kb), and represents a unique tool for analyzing and sequencing large eucaryotic genes, or introducing large DNA fragments in lambda phages or cosmid vectors. Vent polymerase (also known as Tli polymerase since isolated from *Thermococcus litoralis*) has an error rate comprised between that of Taq and Pfu.

Vent polymerase has a half-life of 7 h at 95°C. It is marketed by New England Biolabs, together with a modified version that lacks exonuclease (proofreading) activity. Two DNA polymerase (Pol I and II) that exhibit a 32 '!52 exonuclease proofreading activity were also isolated from *Pyrococcus abyssi* (Gueguen et al. 2001). Pol I DNA polymerase from *P. abyssi* (Pab), marketed by Qbiogen as "Isis DNA Polymerase", has an error rate (0.66×10^{-6}) similar to that of Pfu. However, Pab is more thermostable than Pfu or Taq (Pab has a half-life of 5 h at 100°C), therefore being very useful for conducting PCR reactions that involve high temperature incubations.

Most of DNA polymerases and reverse transcriptases (described below) are commercially available in "Hot Start" (or equivalent) version. Hot Start modification is aimed at preventing non-specific priming events such as template/primer hybridization or primer dimmer formation, which occur in low stringency conditions (during PCR preparation). Non-specific priming events generate secondary products during the preparation and in the first cycle of the PCR, and are further amplified in subsequent cycles.

Hot Start technology introduces a "barrier" between secondary structures and the polymerase DNA-binding site. The "barrier", which can be an antibody, an oligonucleotide, or a reversible chemical modification of the polymerase's amino acids, is released from the enzyme during the first denaturation cycle of the PCR, thus restoring enzymatic activity only after denaturation of secondary structures. Hot Start modifications are very efficient in limiting polymerase activity at room temperature, and thereby facilitate PCR preparation while enhancing PCR specificity.

Reverse Transcriptase (RNA-dependent DNA Polymerase EC 2.7.7.49)

Until the 1960's, the transfer of genetic information was thought to flow unidirectionally from DNA to RNA (Crick 1958). Characterization of reverse transcriptase (RT) from Rous Sarcoma Virus (RSV) (Temin and Mizutani 1970) and Rauscher Leukaemia virus (RLV) (Baltimore 1970) exemplified that single stranded DNA could be synthesized from an RNA template. The resulting single-stranded DNA is a "complementary" copy of the RNA template, or cDNA.

Viral Reverse-transcriptases

RSV and RLV belong to the group of retroviridae (retroviruses). The life cycle of retroviruses includes an RT-directed transcription of their RNA genome and formation of a proviral double stranded molecule of DNA, which is integrated in the host genome. Infectious viral progeny is produced from transcription of the proviral DNA.

Many viruses contain a reverse transcription stage in their replication cycle. Metaviridae are closely related to retroviruses and exist as retrotransposons in the eucaryotic host genome. Retrotransposons are mobile elements that amplify (multiply) through intermediate RNA molecules, which are reverse transcribed and integrated at new places in the host genome. Examples of metaviridae include the Saccharomyces cerevisiae Ty3 virus, Drosophila melanogaster gypsy virus, and Ascaris lumbricoides Tas virus.

Pseudoviridae (*e.g.* Saccharomyces cerevisiae Ty1 virus and Drosophila melanogaster copia virus) have a segmented single stranded RNA genome.

Hepadnaviridae (like the hepatitis B virus (Seeger and Mason 2000)) have a genome made of two uneven strands of DNA that exist as stretches of both single and double stranded circular DNA. The viral polymerase, which catalyses RNA- and DNA-dependent DNA synthesis, possesses both RNase H and protein priming activities. Upon infection, the relaxed circular DNA is converted into a circular DNA that is transcribed by the host RNA polymerase. This "pre-genomic" RNA is retrotranscribed in cDNA by the viral RT to give rise to the genomic viral DNA strands.

Caulimoviridae are unenveloped viruses that infect plants, including cauliflower (cauliflower mosaic virus) and soybean (soybean chlorotic mottle virus). Upon infection, the polyadenylated viral RNA is reverse-transcribed by the viral RT to give rise to genomic double-stranded DNA molecules. Viral RNA is also used for viral protein synthesis in the cytoplasm of infected cells. Although caulimoviridae's life cycle resembles that of retrovirus, replication of caulimoviridae does not involve the integration of proviral DNA in the host genome.

Non-viral Reverse Transcriptases

Transposons Nearly all organisms contain variable amounts of repetitive mobile DNA known as transposable elements (TEs) or transposons. TEs constitute more than 80% of the total genome in plants, while it represents about 45% of the human genome. Finnegan's classification (Finnegan 1989) distinguishes RNA transposons (Class I or retrotransposons), which amplify through a "copy and paste" type of transposition, and DNA transposons (Class II), which use a "cut and paste" type of transposition. However, the newer Wicker's classification takes into account the recent discovery of bacterial and eukaryotic TEs that copy and paste without RNA intermediates, and of new miniature inverted repeat transposable elements (MITEs) (Wicker et al. 2007).Retrotransposons, which make up to 42% of human genome, encode an RT used to make a cDNA copy of an RNA intermediate, which is produced after transcription of the retrotransposon integrated in the genome.

Telomerase Telomeres are non-coding, linear sequences that cap the ends of DNA molecules in eukaryotic chromosomes. Telomeres are made of up to 2,000 repeats of TTAGGG stretches. In normal cells, the DNA replication machinery is unable to duplicate the complete telomeric DNA. As a result, telomeres are shortened after every cell division. Having reached a critical length, telomeres are recognized as double strand break DNA lesions, and cells eventually enter senescence.In embryonic and adult stem cells, which have an extended lifespan, telomere length is maintained through activation of telomerase. Telomerase is a ribonucleoproteic complex that contains an RNA (TElomere RNA Component, or TERC) and an RT (TElomere Reverse Transcriptase, or TERT). The RNA provides the AAUCCC template directing the synthesis of TTAGGG repeats by the RT. Interestingly, telomerase is also activated during carcinogenesis (Raynaud et al. 2008).

Reverse Transcriptase for Molecular Biology

The use of RT in fundamental and applied molecular biology has been propelled by the introduction of the Reverse Transcription Polymerase Chain Reaction (RT-PCR). As a result, commercial sources of RT have flourished over the past two decades.

Because RTs are deprived of 32 '!52 exonuclease (proofreading) activity, they have much lower fidelity than DNA-dependent DNA polymerases. For instance, HIV RT has an error rate of 1 mutation per 1,500 nucleotides, whereas RT of avian and murine origin generate 1 mutation per 17,000 and 30,000 nucleotides, respectively.

AMV/MAV RT

The RT the most commonly used in molecular biology is the one that allows the replication of the Avian Myeoloblastosis Virus (AMV), an alpha retrovirus that induces myeoloblastosis in chicken (Baluda et al. 1983). Interestingly, cloning of the AMV genome identified the v-myb oncogene as responsible for intense proliferation of transformed myeloblasts. Insertion of v-myb oncogenic sequences in the AMV genome interrupts the coding sequence of the RT, thus making AMV RT-deficient. Like every deficient viruses, AMV depends on a helper virus for its replication. AMV helper, the Myeoloblastosis Associated Virus (MAV), is in fact the "real" source of RT in the life cycle of AMV, and for molecular biology (Perbal 2008).

The AMV/MAV RT is composed of two structurally related sub-units designated α and β (65 kDa and 95 kDa, respectively). The α subunit of the enzyme provides RT and RNase H activities. The RT activity requires the presence of a primer and a template. AMV/MAV RT is widely used to copy total messenger RNAs using polydT or random primers. RNase H activity is generated by proteolytic cleavage of the α subunit and associated with a 24 kDa fragment. RNase H is a processive exoribonuclease that degrades specifically RNA strands in RNA— DNA hybrids in either 52 '!32 or 32 '!52 directions.

The use of reverse transcriptase has found many applications in molecular cloning, two well-documented examples being the synthesis of cDNA from RNA in the preparation of expression libraries, as a first step for quantitative PCR, and for nucleotide sequencing. Several protocols have establish optimal conditions for high yields reactions (Berger et al. 1983), or for the synthesis of large RNA templates (Retzel et al. 1980).

MuLV RT

The *pol* gene of Moloney murine leukemia virus (M-MuLV) encodes an RT lacking DNA endonuclease activity, and exhibiting a lower RNAse H activity than AMV/MAV RT (Moelling 1974). M-MuLV RT is 4-times less efficient, and at least 4-times less stable than AMV/MAV RT. Therefore, comparable yields of cDNA synthesis require six to eight times more M-MuLV RT than AMV/MAV RT. However, M-MuLV RT is able to generate longer transcripts than does AMV/MAV RT when used in excess (Houts et al. 1979).

Thermostable Reverse Transcriptases

Several RTs identified in thermophilic organisms are commercially available and are useful in certain challenging conditions (*e.g.* to transcribe templates with high CG contents or abundant secondary structures).

Thermus thermophilus (Tth) is a DNA polymerase that efficiently reverse-transcribes RNA in the presence of $MnCl_2$. Upon chelation of the Mn^{2+} ions, its DNA polymerase activity allows for PCR amplification in the presence of $MgCl_2$. Epicentre's MonsterScript™ RT is an Mg^{2+}-dependent thermostable RT that lacks RNase H activity, and is fully active at temperatures up to 65°C. According to its manufacturer, this enzyme can produce cDNA larger than 15 kb. The Klenow fragment of *Carboxydothermus hydrogenoformans* (*C. therm.*) polymerase is a Mg^{2+}-dependent RT that is active at temperatures up to 72°C. It is marketed by Roche (C. therm. Mix) for RT PCR uses. Thermo-X™ RT from Invitrogen has a half-life of 120 min at 65°C, which is the highest stability reported so far.

RNA Polymerase (DNA Dependent-RNA Polymerases)

DNA-dependent RNA polymerases catalyze the 52 '!32 elongation of RNA copies from DNA templates, a process called transcription. Like DNA polymerases that lack proofreading exonuclease activity, RNA polymerases can add an extra base at the end of a transcript. The two main RNA polymerases used in molecular biology are SP6- and T7- RNA polymerases. SP6 RNA polymerase is a 96 kDa polypeptide purified from SP6 bacteriophage-infected *Salmonella typhimurium* LT2, while T7 RNA polymerase is a 98 kDa polypeptide (Stahl and Zinn 1981) produced by the T7 bacteriophage. RNA polymerases are used for *in vitro* synthesis of anti-sense RNA transcripts (Melton et al. 1984), production of labeled RNA probes, or for RNase protection mapping (Zinn et al. 1983).

SP6 and T7 RNA polymerases are similar enzymes: both require Mg^{2+} and a double-stranded DNA template, and are greatly stimulated by spermidine and serum albumine (Butler and Chamberlin 1982). In addition, T7 RNA polymerase typically requires the presence of dithiothreitol in reaction buffer.

SP6 and T7 RNA polymerases are extremely promoter specific and will transcribe any DNA sequence cloned downstream of their respective promoter (SP6 and T7). Importantly, SP6- or T7- RNA polymerase transcription proceeds through poly(A) stretches (Melton et al. 1984), and thus can progress around a circular template multiple times before disassociating. Therefore, linearization of the template prior to translation (a blunt or 52 overhang is recommended (Schenborn and Mierendorf 1985)) will guaranty an efficient termination of transcription. The RNA produced under these conditions is biologically active (Krieg and Melton 1984) and can be properly spliced (Green et al. 1983).

Ligases

DNA Ligases (EC 6.5.1.1 and 6.5.1.2 for ATP and NAD⁺ DNA Ligases, Repectively): DNA ligases connect DNA fragments by catalyzing the formation of a phosphodiester bond between a 32 -OH and a 52 -phosphate group at a single-strand break in double-stranded DNA (Lehman 1974). In cells, DNA ligases are essential for joining Okazaki fragments during replication, and in the last step of DNA repair process. DNA ligases are used in molecular biology to join DNA fragments with blunt or sticky ends such as those generated by restriction enzyme digestion, add linkers or adaptors to DNA, or repair nicks.

DNA ligases operate in a three-step reaction. The fist step involves the creation of a ligase-adenylate intermediate, in which a phosphoamide bond is created between a lysine residue and one AMP molecule of the enzyme cofactor (ATP or NAD^+). Second, the AMP is transferred to the 52 -phosphate end of the DNA nick to form a DNA-adenylate (AppDNA). Finally, a nucleophilic attack from the 32 end of the DNA nick directed to the AppDNA results in joining of the two polynucleotides and release of AMP. Original observations suggested that bacterial DNA ligases use NAD^+ as a cofactor whereas DNA ligases from eukaryotes, viruses and bacteriophages use ATP. However, it is known now that some ligases can accept either cofactor, even though both cofactors are not equally efficient.

Non-thermostable DNA Ligases

The smallest known DNA ligases are the ATP-dependant DNA ligase from Chlorella virus and bacteriophage T7 (34 and 41 kDa, respectively). They are much smaller than eukaryotic DNA ligases, which can reach 100 kDa in size. While T7 and chlorella-encoded ATP-dependent ligases both contain only a nucleotidyl-transferase domain and an OB-fold domain, the eukaryotic ligases contain additional domains that include zinc fingers and BRCT (C-terminal portion of BRCA-1) domains with nuclear or mitochondrial localization signals.

The ligase the most frequently used in molecular biology is the bacteriophage T4 DNA ligase. T4 DNA ligase is a 68 kDa monomer that requires Mg^{2+} and ATP as cofactors. T4 DNA ligase can connect blunt and cohesive ends, or repair single stranded nicks in duplex DNA, RNA, or DNA/RNA hybrids.

The *E. Coli* DNA ligase works preferentially on cohesive double-stranded DNA ends. However, it is also active on blunt ends DNA in the presence of Ficoll or polyethylene glycol. Hybrids such as DNA-RNA or RNA-RNA are not efficiently formed by *E. Coli* DNA ligase. This can be used as an advantage when double stranded DNA ligation is wanted and blunt end ligation needs to be avoided.

Because DNA ligases' activity depend on several factors (such as temperature, fragment concentration, nature of fragments — blunt or sticky ends, length of sticky end, stability of hydrogen bonded structure —), it is difficult to presume the ideal incubation conditions for a specific ligation. For instance, ligation of fragments generated by Hind III is 10–40 times faster than the one for fragments generated with Sal I (even though both enzymes generate sticky fragments).

For these reasons, the definition of the ligase unit is very specific: one ligation unit is defined by most suppliers as the amount of enzyme that catalyzes 50% ligation of Hind III fragments of lambda DNA in 30 min at 16°C under standard conditions.

Most standard ligation protocols recommend incubating the ligation reaction overnight at 16°C. However, ligation protocols should be empirically optimized for every ligation and according to the amount of DNA present in the reaction (although most suppliers give guidelines according to reaction volume, not DNA concentration).

Typically, successful ligation with T4 DNA ligase have been reported with incubation varying from 10 min at room temperature to 24 h at 4°C. In addition, it is good practice to verify the activity of a ligase preparation being kept at "20°C for a long period of time, performing a periodical ligation test. A typical test for ligation of sticky ends is performed with Hind III-digested ë DNA. For blunt end ligations, the same procedure can be used with Hae III-digested DNA.

Thermostable DNA Ligases

Thermostable DNA ligases can perform ligation of duplex molecules and repair of single stranded nicks at temperatures ranging from 45 to 80°C. They are highly specific and are very well suited for applications that need high stringency ligations. Thermostable DNA ligases are isolated from diverse sources such as *Thermus thermophilus*, *Bacillus stearothermophilus* (Brannigan et al. 1999), *Thermus scotoductus* (Jonsson et al. 1994), and *Rhodothermus marinus*.

Thermostable DNA ligases are usually not a substitute for T4 or *E. Coli* DNA ligases but are used for very specific techniques such as Ligase Chain Reaction (LCR). LCR is a technique used to detect single base mutations: a primer is synthesized in two fragments that cover both sides of a possible mutation. Thermostable ligase will connect the two fragments only if they match exactly to the template sequence. Subsequent PCR reactions will amplify only if the primer is ligated.

RNA Ligases (EC 6.5.1.3)

RNA ligases catalyze the ATP-dependent formation of phosphodiester bonds between 52 -phosphate and 32 -OH termini of single stranded RNA or DNA molecules. In cells, they act mainly to reseal broken RNAs. Like DNA ligases, RNA ligases operate in a three-step reaction. First, the RNA ligase reacts with ATP to form a covalent ligase-(lysyl-N)-AMP intermediate, and pyrophosphate. Then the AMP moiety of the ligase adenylate is transferred to the 52 -phosphate end of the RNA to form an RNA-adenylate intermediate (AppRNA). In the third step, a nucleophilic attack of the 32 -OH end of the RNA on the AppRNA creates a phosphodiester bond, which seals the two RNA ends.

T4 RNA Ligases

The best characterized RNA ligase is the T4-bacteriophage RNA ligase (gp63), which was identified in T4-infected *E. Coli*. T4 RNA ligase is also the most commonly used RNA ligase in molecular biology. T4 RNA ligase 1 is used to ligate single stranded nucleic acids and polynucleotides to RNA molecules, usually to label RNA molecules at the 32 -end for RNA structure analysis, protein binding site mapping, rapid amplification of cDNA ends (RLM-RACE), ligation of oligonucleotide adaptors to cDNA, oligonucleotide synthesis (Kaluz et al. 1995), 52 nucleotide modifications of nucleic acids, and for primer extension for PCR. T4 RNA ligase can also be used to circularize RNA and DNA molecules. This enzyme is used in Ambion's FirstChoice™ RLM-RACE Kit for tagging the 52 ends of mRNA with oligonucleotide adaptor. A second RNA ligase encoded by the bacteriophage T4 has recently been described (Ho and Shuman 2002). T4 RNA ligase 2, also known as T4 Rnl-2 (gp24.1), catalyzes both intramolecular and intermolecular RNA strand ligation. Unlike T4 RNA ligase 1, T4 RNA ligase 2 is much more active joining nicks on double stranded RNA than on joining the ends of single stranded RNA. T4 RNA ligase 2 ligates 32 -OH/52 -phosphate RNA nicks, and can also ligate 32 -OH of RNA to the 52 -phosphate of DNA in a double stranded structure.

A truncated form of T4 RNA ligase 2 (truncated T4 RNL2, also known as RNL2 [1–249]) is commercialized by New England Biolabs. Truncated T4 RNL2, first 249 amino acids of the full length T4 RNA Ligase 2 (Ho and Shuman 2002), is unable to perform the first adenylation step of the ligation reaction. Thus, the enzyme does not require ATP but does need the pre-adenylated substrate to specifically ligates pre-adenylated 52 end of DNA or RNA to 32 end of RNA molecules. Truncated T4 RNL2 reduces background ligation because it selects adenylated primers. This enzyme can be use for optimized linker ligation for the cloning of microRNAs. An RNA ligase coding frame has also been identified in the pnk/pnl gene (ORF 86) from the baculovirus Autographa californica nucleopolyhedrovirus (ACNV) and from the radiation-resistant bacterium Deinococcus *radiodurans* (Dra) (Martins and Shuman 2004b). DraRnl ligates 32 -OH/52 -phosphate RNA nicks that can occur in either duplex RNA or in RNA/DNA hybrids. However, it cannot ligate nicks in DNA molecules, as it requires a ribonucleotidic 32 -OH.

Thermostable RNA Ligases

Thermostable RNA ligases have been isolated from bacteriophages rm378 and *TS2126* that infect the eubactrium *rhodothermus marinus* and bacterium *thermus scotoductus* respectively.

Phosphate Transfer and Removal

The principal characteristics of the enzymes used for phosphate transfer and removal in molecular biology discussed in this section.

Alkaline Phosphatase (EC 3.1.3.1)

Alkaline phosphatase is purified from either *E. Coli* or higher organisms (*e.g.* calf intestine). It is used for removal of 52 -phosphate groups from nucleic acids in order to prevent recircularization of DNA vectors in cloning experiments. Alkaline phosphatase does not hydrolyze phosphodiester bonds.

Although *E. Coli* and calf intestine alkaline phosphatases have different structures (80 and 140 kDa, respectively), both enzymes are zinc and magnesium containing protein (Reid and Wilson 1971). They are inactivated by chelating agents such as EGTA, and by low concentrations of inorganic phosphate. Importantly, inorganic phosphates need to be removed after restriction endonuclease cleavage and prior to incubation with alkaline phosphatase. Inorganic phosphates are certainly removed if cloning strategy involves gel migration and purification of linearized vector. If no purification is made, dialysis of endonuclease-digested DNA is required prior to alkaline phosphatase treatment (ethanol precipitation does not efficiently remove inorganic phosphates). Similarly, when labelling DNA fragments at the 52 -end (*e.g.* ^{32}P labelling), incubation with alkaline phosphatase needs to precede incubation with polynucleotide kinase in the presence of [^{32}P]deoxynucleoside triphosphate.

T4 Polynucleotide Kinase (Polynucleotide 52 -hydroxy Kinase, EC 2.7.1.78)

Polynucleotide kinase catalyzes the transfer of a phosphate group from an ATP molecule to the 52 -OH terminus of a nucleic acid (DNA or RNA, with no size limitations), the exchange of 52 -phosphate groups, or the phosphorylation of 32 -ends of mononucleotides. Encoded by the *pse* T gene of bacteriophage T4 (Depew et al. 1975), the T4 polynucleotide kinase is a tetramer of four identical 33-kDa monomers. Polynucleotide kinase is commonly used for labelling experiments with radiolabeled ATP utilized as a phosphate donor.

Phosphate transfer activity is optimum at 37°C, pH 7.6, in the presence of Mg^{2+} and reducing reagents (DTT or β2-mercaptoethanol), and with a minimum of 1 mM ATP and a 5:1 ratio of ATP over 52 -OH ends (Lillehaug et al. 1976). When substrate concentration is limited, addition of 6% polyethylene glycol (PEG 8000) in the reaction mixture enhances radiolabeling of recessed, protruding, and blunt 52 -termini of DNA (Harrison and Zimmerman 1986).

When radiolabeled ATP is used (*e.g.* ã^{32}P-ATP), polynucleotide kinase generates labeled DNA or RNA molecules with ^{32}P at their 52 ends, by catalyzing either direct phosphorylation of 52 -OH groups generated after alkaline phosphatase digestion, or exchange of the 52 -phosphate groups. This reaction is efficiently used for labelling DNA or RNA strands prior to base specific sequencing. Polynucleotide kinase is also used for mapping of restriction sites, DNA and RNA fingerprinting, and synthesis of substrates for DNA or RNA ligase. In addition to the kinase activity described above, T4 polynucleotide kinase also exhibits a 32 -phosphatase activity (Cameron and Uhlenbeck 1977). Optimum pH for 32 phosphatase activity is comprised between 5 and 6, *i.e.* more acidic than for phosphate transfer activity. Based on this property, protocols using T4 polynucleotide kinase as specific 32 -phosphatase have been developed (Cameron et al. 1978).

A T4 polynucleotide kinase lacking 32 -phosphatase activity has been purified from *E. Coli* infected with a mutant T4 phage producing an altered *pse* T1 gene product (Cameron et al. 1978). Like the wild-type enzyme, the mutant polynucleotide kinase is made of four subunits of 33 kDa each, effectively transfers the gamma phosphate of ATP to the 52 -OH terminus of DNA and RNA. In addition, the mutant and the wild type polynucleotide kinases require similar magnesium ion concentrations, have the same pH optima and are both inhibited by inorganic phosphate. The mutant polynucleotide kinase is very useful when the 32 -exonuclease activity must be avoided (*e.g.* 52 -^{32}P terminal labelling of 32 -CMP in view of 32 end-labelling of RNA species prior to fingerprinting or sequencing).

Tobacco Acid Pyrophosphatase

Tobacco acid pyrophosphatase is used as a first step for labelling mRNAs at their of 52 ends, which are usually capped, in order to generate radiolabeled probes of for RNA sequencing: *in vitro*, ^{32}P-labeling of mRNA 52 terminus requires the elimination of the 7-methylguanosine and 52 -phosphate moieties of the capped end. Tobacco acid pyrophosphatase hydrolyzes the pyrophosphate bond in the cap's triphosphate bridge, generating a 52 -phosphate terminus on the RNA molecule (leading to p7MeG, pp7MeG, and ppN— pN— mRNA). The generated open cap can then be dephosphorylated by alkaline phosphatase, and labeled with T4 polynucleotide kinase using ã^{32}P-ATP.

Nucleases

Nucleases cleave phosphodiester bonds in the nucleic acids backbone. Based on their mode of action, two main classes have been defined: exonucleases are active at the end of nucleic acid molecules, and endonucleases cleave nucleic acids internally. Deoxyribonucleases cleave DNA and generates nicks (point in a double stranded DNA molecule where there is no phosphodiester bond between adjacent nucleotides of one strand, typically through damage or enzyme action), whereas ribonucleases cleave RNA. Nucleases are double edge swords for molecular biologists: on one hand, they are the worst enemy to nucleic acids integrity and, on the other hand, they are very useful to cut and manipulate nucleic acids for cloning purposes.

Deoxyribonucleases

Deoxyribonuclease I (DNase I, EC 3.1.21.1)

Deoxyribonuclease I (DNase I) is an endonuclease that acts on single- or double-stranded DNA (either isolated or incorporated in chromatin). DNAse I is used for nick translation of DNA, for generating random fragments for dideoxy sequencing, to digest DNA in RNA or protein preparations, and for DNA-protein interactions analysis in DNase footprinting.

DNase I is a 31 kDa glycoprotein, usually purified from bovine pancreas as a mixture of four isoenzymes (A, B, C, and D). It is important to note that crude preparations of DNAse I are often contaminated with RNase A. Thus, great attention should be paid to the quality control provided by the manufacturer to ensure lack of RNase A activity in DNase I preparations. At the end of the reaction, DNase I can be removed from the preparation by thermal denaturation at 75°C for 5 min in the presence of 5 mM EGTA (Huang et al. 1996).

DNase I-catalyzed cleavage occurs preferentially in 32 of a pyrimidine (C or T) nucleotide, and generates polynucleotides with free 32 -OH group and a 52 -phosphate. In the presence of Mg^{2+}, DNase I hydrolyzes each strand of duplex DNA independently, generating random cleavages. Maximal activity is obtained in the presence of Ca^{2+}, Mg^{2+}, and Mn^{2+} ions (Kunitz 1950). The nature of the divalent cations present in the incubation mixture affects both specificity and mode of action of DNase I. For instance, in the presence of Mn^{2+}, DNase I cleaves both DNA strands at approximately the same site, producing blunt ends or fragments with 1–2 base overhangs.

Exonuclease III (Exodeoxyribonuclease III, EC 3.1.11.2)

Exonuclease III was first isolated from the BE 257 *E. Coli* strain, which contains a thermo-inducible overproducing plasmid (pSGR3). Exonuclease III is a 28 kDa monomeric enzyme (Weiss 1976) with several interesting activities (Mol et al. 1995) 32 -exonuclease activity: exonuclease III catalyzes the removal of mononucleotides from 32 -OH ends of double-stranted DNA; phosphatase activity: exonuclease III dephosphorylates DNA chains that terminate with a 32 -phosphate group (this type of chain is usually inert as a primer and inhibits DNA polymerase action); iii) RNase H activity: exonuclease III degrades RNA strands of DNA/RNA hybrids (Rogers and Weiss 1980) endonuclease activity: exonuclease III cleaves apurinic and apyrimidic bases from the sugar phosphate backbone in DNA (Keller and Crouch 1972). Another significant feature of exonuclease III is its relative specificity for double-stranted DNA. When this enzyme acts as an endonuclease, it generates a gap at nicks in the double-stranted DNA, whereas when it acts as an exonuclease, it generates a 52 -protruding end (that is resistant to further digestion since exonuclease III is not active on single-stranted DNA). Exonuclease III is unique among the exonucleases in its phospho-monoesterase action on a 32 -phosphate terminus. Exonuclease III is often used in conjunction with the Klenow fragment of *E. Coli* DNA polymerase to generate radio-labeled DNA strands. It is also used sequentially with S1 nuclease to reduce the length of double stranded DNA.

Optimum pH for exonuclease III's endonucleolytic and exonucleolytic activities is between 7.6 and 8.5, while phosphatase activity is maximal at pH between 6.8 and 7.4. The presence of Mg^{2+} or Mn^{2+} ions is required for optimal activity, while the presence of Zn^{2+} inhibits enzyme activity.

Bal 31 Nucleases (EC 3.1.11)

Two distinct molecular species described as fast (F) and slow (S) Bal 31 nucleases have been purified from the culture medium of *Alteromonas espejiana* Bal 31. Both species shorten duplex DNA (at both 32 and 52 ends) without introducing internal nicks, and exhibit a highly specific single stranded DNA endonuclease activity (cleaves at nicks, gaps and single-stranded regions of duplex DNA and RNA).

The purified F-Bal 31 nuclease is used for restriction mapping, removing long stretches (up to thousands of base pairs) or short stretches (tens to hundreds of base pairs) from duplex DNA, mapping B-Z DNA junctions, cleaving DNA at sites of covalent lesions (such as UV-induced), and shortening RNA molecules. The S-Bal 31 is a slower acting enzyme used only for restriction mapping and removal of short stretches from DNA duplex. Since Ca^{2+} is an essential cofactor, EGTA is required to inactivate both enzymes.

Most Bal 31-generated DNA fragments have fully base paired ends, which may further be ligated to any blunt end DNA fragments like molecular linkers. A small fraction of Bal 31-generated DNA fragments is harboring 52 -protruding ends, suggesting that Bal 31 acts sequentially as a 32 '!52 exonuclease, followed by endonucleolytic removal of the protruding ends. Thus, the efficacy of a blunt-end ligation can be increased by utilizing Klenow fragment of *E. Coli* DNA polymerase or T4 DNA polymerase to fill the 52 -protuding ends generated by Bal 31.

Although F and S species of Bal 31 have similar activities on single stranded DNA, they act differently on double stranded DNA: F-Bal 31 can shorten duplex DNA approximately 20 times faster than S-Bal 31, although the reaction rate of digestion is dependent upon the C/G content of the substrate DNA. When tested under standard conditions, 1 ìg/ml of the F- and S-Bal 31 shortens DNA at rates of approximately 130 and 10 base pairs/terminus/ minute, respectively.

Exonuclease VII (EC 3.1.11.6)

Exonuclease VII has first been isolated from the *E. Coli* K12 strain as a 88 kDa polypeptide. Exonuclease VII specifically degrades single stranded DNA, with no apparent activity on RNA or DNA/RNA hybrids. Since it is able to attack either end of the DNA molecules, it is often utilized to degrade long single protruding strands from duplex DNA (such as those generated by restriction endonucleases) and generate blunt end DNA. Exonuclease VII differs from exonucleases I and III in that i) it can degrade DNA from either its 52 or 32 end, ii) it can generate oligonucleotides, and iii) it is still fully active in 8 mM EDTA. Exonuclease VII is particularly useful for rapid removal of single-stranded oligonucleotide primers from a completed PCR reaction, when different primers are required for subsequent PCR reactions (Li et al. 1991).

Ribonucleases

Pancreatic Ribonuclease (RNase A, EC 3.1.27.5)

Pancreatic ribonuclease (also called RNase A) is an endonuclease that cleaves single stranded RNA at nucleoside 32 -phosphates and 32 -phospho-oligonucleotides ending in Cp or Up. RNase A is used to reduce RNA contamination in plasmid DNA preparations, and for mapping mutations in DNA or RNA by mismatch cleavage, since it cleaves the RNA in RNA/DNA hybrids at sites of single nucleotide mismatch.

RNAse A degrades the RNA into 32 -phosphorylated mononucleotides and oligonucleotides, generating a 22,32 -cyclic phosphate intermediate during the reaction.

RNase A is active under very different conditions and difficult to inactivate. Hence, great care needs to be taken to avoid contamination with RNAse A and further degradation of RNA samples. RNase-inhibitor from human placenta might be used to inhibit RNAse activity. However, we recommend using diethyl pyrocarbonate (DEPC), guanidinium salts, beta-mercaptoethanol, heavy metals, or vanadyl-ribonucleoside-complexes for efficient inhibition of RNAse A.

RNase A is available from many commercial sources. As stated above for other enzymes, very few suppliers provide useful information regarding origin or purity of their enzyme preparation.

Ribonuclease H (3.1.26.4)

Ribonuclease H (RNase H) specifically degrades RNA in RNA/DNA hybrids. It allows removal of RNA probes from prior hybridizations, or removal of poly-A tails at the 32 end of mRNAs.

Optimal activity of RNase H is achieved at a pH comprised between 7.5 and 9.1 (Berkower et al. 1973) in the presence of reducing reagents. RNase H activity is inhibited in the presence of N-ethylmaleimide (a chemical that react with SH groups), and is not markedly affected by high ionic strength (50% activity is retained in the presence of 0.3 M NaCl). RNase H requires Mg^{2+} ions, which can be replaced partially by Mn^{2+} ions.

Other Ribonucleases

Many other ribonucleases have been used for RNA sequencing. Each of them has specific cleavage requirement and specificity. Below is a list of these ribonucleases.

Ribonuclease Phy I can be used for rapid sequencing of RNA. It is isolated from cultures of *Physarum polycephalum*. Ribonuclease Phy I cleaves the RNA molecule at G, A, and U, but not at C residues. The products are 32 mononucleotides with 52 C termini.

RNase CL3 is used for sequencing RNA. It is isolated from chicken liver (*Gallus gallus*). RNase CL3 digests RNA adjacent to cytidilic acid in a ratio of 60 C residues digested for every U residue digested. The enzyme activity is inhibited by poly-A tracts, unless spermidine is added. Used primarily for RNA sequencing (Lockard et al. 1978), Cereus ribonuclease is an endoribonuclease that preferentially cleaves RNA at U and C residues.

Ribonuclease Phy M, purified from Physarum polycephalum, has also been used for sequencing of RNA (Donis-Keller 1980). Ribonuclease Phy M preferentially cleaves RNA at U and A residues. Occasionally, cleavage at G may occur. This unwanted reaction can be minimized in the presence of 7 M urea.

RNase T1 is an endoribonuclease that specifically degrades single-stranded RNA at G residues. It cleaves the phosphodiester bond between 32 -guanylic residues and the 52 -OH residues of adjacent nucleotides with the formation of corresponding intermediate 22, 32 -cyclic phosphates. RNase T1 is an 11 kDa protein purified from *Aspergillus oryzae.*

RNAse T2 is also purified from *Aspergillus oryzae.* It has a molecular weight of 36,000, and cleaves all phosphodiester bonds in RNA, with a preference for adenylic bonds. RNase U2, purified from *Ustilago sphaerogena,* is utilized as complement of RNase T1 in RNA sequencing, to discriminate purines residues since it specifically cleaves adenine residues when incubated at 50°C, pH3.5, and in the presence of 8 M urea. When incubated under standard conditions (Takahashi 1961), RNase U2 specifically cleaves the 32 - phosphodiester bond adjacent to purines, therefore generating purines 32 - phosphates or oligonucleotides with purine 32 -phosphate terminal groups. Cyclic 22,32 purine nucleotides are obtained as intermediates and that reversal of the final reaction step can be used to synthetize ApN and GpN. RNase U2 is thermostable (80°C for 4 min) in aqueous solution at pH 6.9.

DNA/RNA Nucleases

S1 Nuclease (EC 3.1.30.1)

S1 nuclease is a very useful tool for measuring the extent of hybridization (DNA— DNA or DNA— RNA), probing duplex DNA regions, and removing single stranded DNA of protruding ends generated by restriction enzymes. S1 nuclease is purified from *Aspergillus oryzae.* This enzyme degrades RNA or single stranded DNA into 52 mononucleotides, but does not degrade duplex DNA or RNA— DNA hybrids in native conformation. S1 nuclease is a 32 kDa metalloprotein (Vogt 1973). It requires Zn^{2+} for activity. Co^{2+} and Hg^{2+} can replace Zn^{2+}, but are less effective as cofactors. S1 nuclease's optimal activity is achieved at pH 4.0–4.3 (a 50% reduction of activity is observed at pH 4.9, and the enzyme is inactive at pH>6.0). S1 nuclease is strongly inhibited by chelating agents such as EDTA and citrate, or by low concentration of sodium phosphate (as low as 10 mM). Nuclease S1 is resistant to denaturing agents such as urea, SDS, or formamide and is thermostable (Ando 1966).

S1 nuclease hydrolyzes single-stranded DNA five times faster than RNA, and 75,000 times faster than double-stranded DNA. The low level of strand breaks which can be introduced by S1 nuclease in duplex DNA can be further reduced by high salt concentration (*i.e.* 0.2 M). S1 nuclease is active at S1 sensitive sites generated by negative supercoiling of the helical DNA structure, UV irradiation, or depurination (Shishido and Ando 1975).

S1 nuclease is widely used when specific removal of single portions of duplex DNA, RNA/DNA hybrids, or RNA molecules is required. Among these applications are mapping of spliced RNA molecules, isolation of duplex regions in single stranded viral genomes, probing strand breaks in duplex DNA molecules, cleavage of regions with lesser helix stability, localization of inverted repeated sequences, introduction of deletion mutation at D loop sites in duplex DNA, and mapping of the genomic regions involved in interactions with DNA binding proteins (Meyer et al. 1980).

Mung bean Nuclease

Mung bean nuclease can be used in the same way as nuclease S1, to remove protruding ends in duplex DNA or for transcription promoter mapping. Mung bean nuclease is a single stranded specific DNA and RNA endonuclease purified from mung bean sprouts. It yields 52 -phosphate terminated mono- and oligonucleotides. Complete duplex DNA degradation may occur when high enzyme concentration and extended incubation time are used. This is due to a two-step process: the enzyme first introduces single stranded nicks, followed by double stranded scissions and exonucleolytic digestion of the resulting fragments.

Proper trimming of the 52 -protruding extensions is achieved when the final blunt end contained a G-C base pair at its terminus (Ghangas and Wu 1975). Presence of an A-T base pair at the position where the fragment would end after trimming seemed to interfere with precise removal of the protruding end. Nucleotide composition of the overhang does not affect efficiency or quality of the nucleolytic digestion (Ghangas and Wu 1975). Mung bean nuclease was also proven to be very useful for excising cloned DNA fragments inserted in vectors following a dA·dT tailing (Wensink et al. 1974). Because poly(dA/dT) are not recognized as typical double stranded structures (Johnson and Laskowski 1970), they are hydrolyzed at half the rate of single stranded tails, but they are more efficiently cleaved than other duplex regions in DNA.

Mung bean nuclease requires Zn^{2+} and a reducing agent such as cysteine for maximum activity and stability. It is inhibited by high salt concentrations (80 to 90% inhibition in 200–400 mM NaCl). Triton X-100 (0.001% w/v) can increase mung bean nuclease stability (when used at low concentrations such as less than 50 U/ìl) and to prevent nuclease adhesion to surfaces.

Methylases

Methylases from Bacterial Restriction-modification Systems

Bacteria use restriction endonucleases to degrade foreign DNA introduced by infectious agents such as bacteriophages. In order to prevent destruction of its own DNA by the restriction enzymes, bacteria mark their own DNA by adding methyl groups to it (methylation). This modification, which must not interfere with the DNA base-pairing, usually affects only a few specific bases on each strand. Methylases are part of the bacterial restriction- modification (RM) system.

They catalyze the transfer of methyl groups from S-adenosyl-methionine (SAM) to specific nucleotides of double stranded DNA molecules. Methylases from type II RM systems (the most common) are encoded by separate proteins and act independently of their respective restriction endonucleases, whereas methylase and restriction activities of type I and III RM systems are provided by a unique protein complex (Wilson 1991).

Methylations often inhibit restriction enzymes that recognize the corresponding sequences (Sistla and Rao 2004), although there are exceptions to this rule (Gruenbaum et al. 1981a) some restriction endonucleases cleave DNA at a recognition sequence being modified by the *dam* or *dcm* methylases some bacteria have restriction endonucleases that only degrade methylated DNA and not the host unmethylated DNA (to overcome the camouflage evolution of some bacteriophages that contain methylated DNA).

Methylases are often used in molecular cloning to protect DNA from digestion, during DNA cloning or cDNA or genomic library construction. For example, when cloning a DNA fragment at the BamH I site of a given vector, the BamHI methylase can first be used to protect potential internal Bam HI sites in the vector, before addition of Bam HI linkers and digestion with the Bam HI endonuclease.

A methylase can also inhibit a restriction endonuclease of different RM systems (*e.g.* TaqI methylase inhibits BamHI restriction enzyme). The extent of overlap between restriction and methylation sequences determines the extent to which methylation alters endonuclease cleavage specificity and/or activity.

There are two types of overlaps: in the first case, the methylation sequence completely overlaps with the one of restriction endonuclease, and methylation alters cleavage specificity. In the second case, the methylation sequence partially overlaps with the one of restriction endonucleases, and nucleotide methylation modifies the endonuclease recognition sequence, which becomes resistant to the endonuclease activity.

The first class of overlap occurs when restriction endonucleases have degenerated recognition sequence, and when methylases are active on only one of the possible sites. For example, the Hinc II recognition sequence is GTPyPuAC, which means that Hinc II can cleave duplex DNA at the following four combinations: GTCGAC, GTCAAC, GTTGAC, and GTTAAC.

TaqI methylase (designated M.TaqI) catalyzes the transfer of a methyl group to the adenine residue of the TCGA sequence. Thus, among the four possible Hinc II recognition sites, only those containing a TCGA sequence are methylated by M.TaqI, and consequently become resistant to Hinc II digestion. This kind of overlap can be used to create new specificities for restriction endonucleases in duplex DNA (Nelson et al. 1984).

The second class of overlap occurs at the boundaries of the recognition sequences for a restriction endonuclease and a methylase. For example, when the GGATCC sequence of a Bam HI restriction site is followed by GG, the Bam HI site partially overlaps with the CCGG methylation site of M.MspI. Since M.MspI transfers a methyl group to the 52 cytosine residue in the CCGG sequence, it methylates the Bam HI site at its internal cytosine, thereby making this sequence resistant to the Bam HI endonuclease.

Dam and Dcm Methylases

Most B, K, and W strains of *E. Coli* also contain two site-specific DNA methylases, which are not part of the RM system and are encoded by the *dam* and *dcm* genes (Pirrotta 1976). The *dam* methylase transfers a methyl group from SAM to the N6 position of the adenine residue in the sequence GATC, while the *dcm* methylase (also known as *mec*) transfers a methyl group from SAM to the internal cytosine residues in the sequences CCAGG or CCTGG.

Dam methylation regulates post-replication mismatch repairs. All DNA isolated from *E. Coli* is not methylated to the same extent. For instance, the pBR322 plasmid DNA purified from *E. Coli* after amplification with chloramphenicol is resistant to the digestion by Mbo I, which does not cut at methylated sites. Furthermore, only 50% of the ë DNA obtained from a *dam*$^{+}$ strain of *E. Coli* is methylated. The degree of *in vivo* cytosine methylation in the sequence CpG is related to the level of gene expression in eukaryotic cells, and quantity of methylated cytosine residues inversely correlates with gene activity.

Practical Considerations on the in Vitro use of Methylases

DNA fragments obtained by restriction endonuclease digestion of *in vitro* methylated DNA are in most respects indistinguishable from unmethylated DNA. However, it is important to note that methylated cytosine does not generate a band in the C channel when sequencing DNA by the Maxam and Gilbert method. In addition, the presence of methyl-cytosine residues in DNA reduces efficiency of transformation in most common strains of *E. Coli*.

This second problem results from the existence of restrictions systems in *E. Coli* K12 responsible for a specific degradation of DNA containing methylated cytosines. These restriction systems, designated 5-methylcytosine-specific restriction enzyme *(mcr)-A* and *mcrB*, are similar to those degrading hydroxymethylcytosine-containing DNA (*rglA* and *rglB*). The *mcrA*$^+$ strains cleaves DNA modified by the Hpa II methylase while *mcrB* cleaves DNA modified by the Hae III, Alu I, Hha I, and Msp I methylases. *E. Coli* strains deficient in the *mcrB* system are available from New England Biolabs. DNA methylases from of type II RM systems perform the methylation reaction under similar conditions as do restriction endonucleases, except that the methylase requires SAM as a methyl group donor. Therefore, it is generally acceptable to carry out the methylation reaction using standard restriction endonuclease buffers to which SAM has been added. Importantly, methylases do not require divalent cations to be active, while most restriction endonucleases do.

Other Enzymes

Polyadenylate Polymerase of E. Coli *(Polynucleotide Adenylyl-transferase, EC 2.7.7.19)*

Polyadenylate (polyA) polymerase of *E. Coli* is a 58 kDa polypeptide that polymerizes adenylate residues at the 32 end of various polyribonucleotides, in a template-independent manner. *In vitro*, it is used for RNA 32 extension to prepare a priming site for cDNA synthesis using oligo-dT, or to prepare RNA for poly-dT-based purification. Depending on the experimental conditions, polyA polymerase can polymerize a polyA stretches from 20- to 2,000-nucleotide long.

E. Coli polyA polymerase catalyzes the addition of adenosine monophosphates (AMPs) using ATP as a substrate. ADP, dATP, and GTP are not polyA polymerase substrates, but CTP and UTP are (though at less than 5% of the rate obtained with ATP). PolyA polymerase is inhibited by phosphate and pyrophosphate ions, and aurin tricarboxylic acid, an inhibitor of both RNA polymerase and of the binding of mRNA to ribosomes. PolyA polymerase is a rather unusual enzyme in that its optimal activity requires the presence of high concentrations of monovalent cations (*e.g.* 400 mM NaCl). It is stimulated by Mn^{2+} and is insensitive to antibiotics such as rifampicin and streptolygdin (transcriptional inhibitors of initiation and elongation, respectively). In contrast to polynucleotide phosphorylase, another template-independent polynucleotide synthesizing enzyme, polyA polymerase does not degrade its own polyA product.

PolyA polymerase uses a wide variety of single stranded RNA species as primers. Double stranded RNA and some polynucleotides (such as polyUG, polyC, or di- and trinucleotides) are very poor primers. DNA does not function as a primer. This latter property is common to virtually all other polyA polymerases, except for the enzyme isolated from plants such as maize that shows considerable activity with natural and synthetic oligo- and poly-dNTPs.

Topoisomerase I (EC 5.99.1.2)

Topoisomerase I is a 105 kDa protein (Liu and Miller 1981) that catalyzes the relaxation of supercoiled DNA by transiently cleaving one strain of duplex DNA in the sugar— phosphate backbone, and further ligating the two generated ends. Topoisomerase I is an ubiquitous nuclear protein that is used by cells during processes such as replication, recombinaison, or DNA- protein interactions that involve unwinding of the DNA, and thereby create excessive supercoiling upstream of the unwinding point (Champoux 2001).

In vitro, topoisomerase I was initially used for circular DNA preparation (Martin et al. 1983), and for studying nucleosome assembly or DNA tertiary structures. Later, it was shown that topoisomerase I also catalyzes the covalent transfer of a single stranded DNA (donor) to an heterologous DNA (acceptor). This reaction involves the initial binding of the enzyme to single stranded DNA and formation of a covalent DNA/enzyme complex, which is then able to ligate the single stranded fragment to the 52 -OH end of the acceptor DNA. This ability of topoisomerase I to function as both a nuclease and a ligase has been utilized to develop new cloning vectors (Invitrogen TOPO® Cloning Vectors). The TOPO® cloning method uses topoisomerase I of *vaccinia* virus that i) specifically cuts double stranded DNA at thc end of a (C/T)CCTT sequence, and ii) binds covalently to the 32 -phosphate of the thymidine of the generated DNA. The ligase activity of topoisomerase I can then join an acceptor DNA with compatible ends. TOPO® vectors form Invitrogen are provided as linear duplex DNA containing a topoisomerase I covalently linked to the 32 -phosphate on one strand, and a GTGG overhang on the other strand. Cloning is achieved using PCR products that have been amplified using a forward primer that contains four extra bases (CACC) at the 52 end. The overhang in the cloning vector (GTGG) hybridizes the 52 end of the PCR product, anneals to the added bases, and topoisomerase I ligates the PCR product in the correct orientation. Topoisomerase I also ligates the PCR product at the blunt 32 end. Topoisomerase I is active at pH 7.5 in the presence of 50 to 200 mM NaCl, and is inactivated by 0.2% SDS. Covalent binding to DNA is stimulated by 5–10 mM Mg^{2+}.

Topoisomerase II (E. Coli DNA gyrase, EC 5.99.1.3)

Topoisomerase II, which is purified from *M. Luteus* (Klevan and Wang 1980), catalyzes the breakage and resealing of both strands of duplex DNA, thereby changing the linking number of discrete supercoiled forms of DNA by two. Topoisomerase II is also capable of reversibly knotting intact circular DNA. Type II topoisomerases are multimeric proteins that require ATP to be fully active. Type II topoisomerase is inhibited by etoposide, a chemotherapeutic agent used to reduce growth of rapidly dividing cancer cells (Baldwin and Osheroff 2005).

Guanylyl Transferase from Vaccinia Virus

Guanylyl transferase is a capping enzyme complex isolated from *Vaccinia* virus. Guanylyl transferase is used to label either 52 di- and triphosphate ends of RNA molecules, or capped 52 ends of RNA after chemical removal of the terminal 7-methyl-guanosine (m^7G) residue. This complex has three enzymatic activities: 1) it acts as an RNA triphosphatase by catalizing the cleavage of a pyrophosphate from a trisphosphate end of an RNA molecule, releasing a bisphosphate RNA end; 2) it is a guanylyl transferase, able to transfer one molecule of GTP to RNA, releasing pyrophosphate (because the 52 end of the RNA molecule ends in a phosphate group, the bond formed between the RNA and the GTP molecule is an unusual 52 -52 triphosphate linkage, instead of the 32 -52 bounds that exist between the other nucleotides forming an RNA strand); and 3) it is an RNA (guanine-7)-methyl transferase that transfers a methyl group from SAM to the 52 guanidine residue of an RNA molecule.

Guanylyl transferase complex does not accept monophosphate RNA as a substrate. Consequently, degraded or nicked RNA will not be labeled except at the 52 -cap end. The optimal amount of enzyme to be used must be determined empirically. During labelling experiments, it is recommended to verify that all labelling is incorporated in an authentic cap structure. To this end, tobacco acid pyrophosphatase can be used on an aliquot of the labeled samples: all labelling should be released in the form of 52 GMP after digestion.

3

Respiration and Lipid Metabolism

Although it is possible to measure respiration by intact tissues it is often useful to be able to isolate uncontaminated, intact and functional mitochondria. Here, we will consider some of the principles for isolating cell organelles. This discussion is not meant to be exhaustive nor will it be a cookbook with recipes to be followed step-by-step.

Choice of Material

If the aim of the experiment is to investigate a specific physiological problem, then the plant material is often well-defined. However, in many cases there is some flexibility in the choice of material and it is therefore wise to follow certain rules of thumb:

- Choose a tissue that is easy to obtain in relatively large amounts (at least 100–200 g fresh weight), such as tubers or hypocotyls.
- Choose a tissue for which a good protocol is available. It can take quite a lot of time and effort to optimize an isolation protocol.

Note that the isolation of mitochondria from green tissues presents special problems because of the large amount of chloroplasts. Roots are a notoriously difficult tissue for isolation of mitochondria.

Disruption of the Tissue

The aim is to rupture as many of the cells in the tissue as possible in order to release their cytoplasm with the mitochondria. Whereas many mammalian tissues are easy to homogenize, plant tissues are much tougher mainly because of the cell walls. There are a number of standard homogenizers, including mixers with rotating knives or razor blades, juice extractors, ball mills, and mortar and pestle; the most suitable method for the specific plant material should be chosen. This may involve some initial optimization experiments.

Homogenization Medium

The medium for homogenizing rat liver only contains an osmoticum and a buffer. That will not work with plant tissues because of the disruption of the large central vacuole (absent in mammalian cells), which releases its fairly acidic content often containing phenols and other potentially harmful substances.

Let us consider the components of the optimal homogenization medium for isolating intact and functional organelles from plant material, one by one.

Osmoticum. This is required to prevent the mitochondria from losing their matrix content because of swelling and bursting. For the isolation of plant mitochondria, there is a tradition of using sucrose or mannitol at 0.2–0.5 M; whereas 0.15 M KCl is standard for the isolation of mammalian mitochondria. In neither case does this mimic precisely the conditions in the cytosol, but it works.

Buffer. The pH in the cytosol of plant cells is around 7.5, and the homogenate should be buffered at pH 7–8. A buffer should therefore be included and a plethora of different buffers are available. Choose an inexpensive one with a pKa of 7–8 (e.g., 4-morpholinopropane sulfonic acid = MOPS) and be sure to use enough to counterbalance the acidic vacuolar content. The tissue type, the tissue/medium ratio, and the buffer concentration determine whether the acidic vacuolar content is properly neutralized. Obviously, special tissues like lemons or CAM plants harvested late at night or in the early morning, require special treatment! Metal ion chelators. EDTA is usually included to bind mainly Ca^{2+} and Mg^{2+}, which may otherwise activate proteolytic enzymes.

Antioxidants. Antioxidants (reductants) such as ascorbic acid, dithiothreitol, or cysteine are used mainly to protect sulfhydryl groups on proteins against oxidation.

Bovine serum albumin (BSA). This protein from the blood of cows is usually included to protect the mitochondria: BSA binds fatty acids (its function in the blood) released by the action of phospholipases on phospholipids. BSA may also bind phenolics that are released from the vacuole. Both of these groups of compounds can otherwise be troublesome later. In addition, BSA may prevent proteases from degrading mitochondrial proteins by serving as an alternative substrate present at high concentrations.

Polyvinylpyrrolidone (PVP). A polymer that is often included, especially when using leaf material, to adsorb phenolic compounds.

Filtration

To remove larger tissue fragments, cell walls, and often starch grains, the homogenate is always filtered, usually through a nylon net with a mesh size of 30–100 μm.

Purification of Mitochondria

The first step of the purification is usually differential centrifugation. This means that the filtered homogenate is centrifuged first at low speed for a short time (e.g., 3000 g for 5–10 min.) and the pellet (containing cell wall fragments, starch grains, nuclei, intact plastids) is discarded. The supernatant is then centrifuged faster and for a longer time (e.g., 10,000 g for 10–20 min.) and the pellet (termed *crude mitochondria*) is collected. It contains intact and damaged mitochondria as well as peroxisomes and plastid membranes. The supernatant is discarded.

As soon as the mitochondria have been pelleted and thus removed from potentially damaging soluble substances in the homogenate, the medium can be greatly simplified. Usually the so-called wash medium contains only osmoticum and weak buffer (no need for a strong buffer since there is no vacuole content to neutralize). However, it may be wise to include EDTA to bind divalent cations as well as BSA to adsorb any fatty acids released by the action of phospholipases. The resuspended crude mitochondria are often diluted to 50–100 ml with wash medium and subjected to another round of slow and fast centrifugation to remove more contaminants.

The aim of the next step is to separate the intact mitochondria from the contaminants. This usually involves a density gradient centrifugation with Percoll either alone (e.g., Struglics et al. 1993) or together with other dense materials (e.g., Day et al. 1985). Percoll (colloidal silica coated with PVP) is inert and osmotically inactive. It has the further advantage that a Percoll gradient is self-generating because the Percoll microspheres pellet very slowly during the centrifugation. The crude mitochondria are layered on top of the wash medium containing the appropriate concentration of Percoll (usually 25–30%). If the Percoll concentration and the speed and duration of the centrifugation have been chosen correctly, the intact mitochondria will form a distinct band, well away from the contaminants. This band can be collected, washed free of Percoll by repeated dilution and centrifugation, and either used immediately, or frozen with 5% (v/v) dimethylsulfoxide (to preserve integrity) in liquid nitrogen and stored at -80°C until use. This preparation can be stable for months, retaining the ability to form a proton electrochemical gradient over the inner membrane.

The yield is 1–50 mg mitochondrial protein per 100 g fresh weight, depending on the tissue. This represents only a small fraction of the total amount of mitochondria in the tissue. The purification of mitochondria from green leaves presents special problems and a good protocol is described in Day et al. (1985). There is, at present, no good protocol for the isolation of both pure and functional mitochondria from Arabidopsis leaves (Hausmann et al. 2003; Keech et al. 2005), but a protocol for the purification of mitochondria from Arabidopsis cell cultures is given in Kruft et al. (2001). Finally, Boutry et al. (1984) demonstrate how purified mitochondria can be isolated from small amounts of plant tissue (down to 0.5 g fresh weight).

Mitochondria are rod-shaped organelles that can be considered the power generators of the cell, converting oxygen and nutrients into adenosine triphosphate (ATP). ATP is the chemical energy "currency" of the cell that powers the cell's metabolic activities.

This process is called aerobic respiration and is the reason animals breathe oxygen. Without mitochondria (singular, mitochondrion), higher animals would likely not exist because their cells would only be able to obtain energy from anaerobic respiration (in the absence of oxygen), a process much less efficient than aerobic respiration. In fact, mitochondria enable cells to produce 15 times more ATP than they could otherwise, and complex animals, like humans, need large amounts of energy in order to survive.

The number of mitochondria present in a cell depends upon the metabolic requirements of that cell, and may range from a single large mitochondrion to thousands of the organelles. Mitochondria, which are found in nearly all eukaryotes, including plants, animals, fungi, and protists, are large enough to be observed with a light microscope and were first discovered in the 1800s.

The name of the organelles was coined to reflect the way they looked to the first scientists to observe them, stemming from the Greek words for "thread" and "granule." For many years after their discovery, mitochondria were commonly believed to transmit hereditary information. It was not until the mid-1950s when a method for isolating the organelles intact was developed that the modern understanding of mitochondrial function was worked out.

The elaborate structure of a mitochondrion is very important to the functioning of the. Two specialized membranes encircle each mitochondrion present in a cell, dividing the organelle into a narrow intermembrane space and a much larger internal matrix, each of which contains highly specialized proteins. The outer membrane of a mitochondrion contains many channels formed by the protein porin and acts like a sieve, filtering out molecules that are too big. Similarly, the inner membrane, which is highly convoluted so that a large number of infoldings called cristae are formed, also allows only certain molecules to pass through it and is much more selective than the outer membrane. To make certain that only those materials essential to the matrix are allowed into it, the inner membrane utilizes a group of transport proteins that will only transport the correct molecules. Together, the various compartments of a mitochondrion are able to work in harmony to generate ATP in a complex multi-step process.

Mitochondria are generally oblong organelles, which range in size between 1 and 10 micrometers in length, and occur in numbers that directly correlate with the cell's level of metabolic activity. The organelles are quite flexible, however, and time-lapse studies of living cells have demonstrated that mitochondria change shape rapidly and move about in the cell almost constantly.

Movements of the organelles appear to be linked in some way to the microtubules present in the cell, and are probably transported along the network with motor proteins. Consequently, mitochondria may be organized into lengthy travelling chains, packed tightly into relatively stable groups, or appear in many other formations based upon the particular needs of the cell and the characteristics of its microtubular network.

The mitochondrion is different from most other organelles because it has its own circular DNA (similar to the DNA of prokaryotes) and reproduces independently of the cell in which it is found; an apparent case of endosymbiosis.

Scientists hypothesize that millions of years ago small, free-living prokaryotes were engulfed, but not consumed, by larger prokaryotes, perhaps because they were able to resist the digestive enzymes of the host organism.

The two organisms developed a symbiotic relationship over time, the larger organism providing the smaller with ample nutrients and the smaller organism providing ATP molecules to the larger one. Eventually, according to this view, the larger organism developed into the eukaryotic cell and the smaller organism into the mitochondrion.

Mitochondrial DNA is localized to the matrix, which also contains a host of enzymes, as well as ribosomes for protein synthesis. Many of the critical metabolic steps of cellular respiration are catalysed by enzymes that are able to diffuse through the mitochondrial matrix. The other proteins involved in respiration, including the enzyme that generates ATP, are embedded within the mitochondrial inner membrane. Infolding of the cristae dramatically increases the surface area available for hosting the enzymes responsible for cellular respiration.

Methods for Establishing Purity, Intactness, and Functionality of the Fractions

It is often important to determine the degree of purity of the mitochondria and their subfractions.

The intactness of the mitochondria is usually established with enzyme assays that take advantage of the fact that an intact membrane is not permeable to certain molecules.

For instance, the outer mitochondrial membrane is impermeable to cytochrome *c*, a 12.5 kDa protein that is too large to pass through the outer membrane pores, and NADH cannot pass through an intact inner membrane.

Functionality is usually tested using an oxygen electrode. If the mitochondria show high rates of oxygen consumption, high respiratory control ratios (>3), and ADP:O ratios close to the theoretical values they are in very good shape. Note that, for mitochondria with an appreciable alternative oxidase activity (measured in the presence of the cytochrome oxidase inhibitor, KCN), the respiratory control ratio and ADP:O ratio should be measured in the presence of an alternative oxidase inhibitor such as salicylhydroxamic acid.

Mitochondrial Activities can be Determined in Situ

It is relatively straightforward to measure the activity of enzymes in the soluble matrix fraction isolated from mitochondria.

However, this fraction is very diluted compared to the matrix inside intact mitochondria (300–500 mg protein/ml!) and protein–protein interactions may have been lost. When measuring the activity of enzymes in fractions containing membranes—intact mitochondria and submitochondrial particles—a detergent can often be used to remove the permeability barrier that the inner membrane constitutes. However, it is also possible to use the ionophore alamethicin to permeabilize the inner membrane to molecules <1000 Da, which makes it possible to measure the activity of matrix and inner membrane enzymes in their native proteinaceous environment. Using alamethicin, it is even possible to measure the activity of cytosolic and mitochondrial enzymes inside intact cells since the ionophore also permeabilizes the plasma membrane.

The Q-Cycle Explains How Protons Can Be Pumped across the Inner Mitochondrial Membrane

The electron transport chain (ETC) in the mitochondrial inner membrane of most eukaryotes consists of four complexes, complex I to complex IV. Complexes I, III and IV pump protons over the inner mitochondrial membrane when transporting electrons.

For each electron pair (corresponding to one NADH or one oxygen atom), four protons are pumped by complex I, four by complex III, and two by complex IV. The detailed three-dimensional structures have been determined for the mammalian complexes III and IV, but not for complex I. The structural information has improved the understanding of the electron paths of complexes III and IV, but the proton-pumping function of the latter is still not fully understood. For complex I, neither electron nor proton paths are known in any detail.

Ubiquinone (Q), also called coenzyme Q, is a mobile electron and proton carrier in the inner mitochondrial membrane, structurally and functionally similar to plastoquinone in the thylakoid membrane of plastids (Figure 1). The ability of Q to carry both electrons and protons—and that it can accept and donate them step-wise—is utilized in electron transport and proton pumping (Figure 2) (Nicholls and Ferguson 2002).

To be reduced, it can receive two electrons from complex I or succinate dehydrogenase and take up two protons from the matrix, forming QH_2. However, at the oxidizing side (complex III), there is actually both oxidation and reduction taking place. A QH_2 will be oxidized to Q at the outer Q-site, releasing two electrons. One of these electrons goes to cytochrome *c*, whereas the other is recycled to reduce an oxidized Q in the inner Q-site—one step to becoming a semiquinone.

The next time a QH_2 is oxidized at the outer Q-site, the semiquinone in the inner site will receive a second electron, becoming a QH_2. So, from two QH_2, we have made one Q and one QH_2, which again can be oxidized. In each cycle (i.e., for each electron passing to cytochrome *c*), two protons are taken up from the matrix and two protons released to the intermembrane space. So, the Q-cycle mediates the pumping of an extra proton per electron transported, increasing energy conservation.

If we look at the whole reaction from succinate to cytochrome *c*, we see that for each succinate, four protons are transported from the matrix to the intermembrane space.

However, at the same time, the two electrons are also transported out. So, this corresponds to the net pumping out of two positively charged species.

Since, energetically, the mitochondrial membrane potential (which determines the energetic cost of transporting a positive charge out) is much larger than the pH gradient (which determines the cost for electroneutral export of a proton), the complex III proton pumping is less energy-demanding than, for example, complex I, which pumps four protons in their completeness.

Multiple Energy-Conservation Bypasses in Oxidative Phosphorylation of Plant Mitochondria

The electron transport chain (ETC) in the mitochondrial inner membrane of most eukaryotes consists of four large protein complexes, complex I to complex IV. The electron transport activity of complexes I, III and IV (the latter two, often jointly called the cytochrome pathway) is coupled to extrusion of protons across the inner mitochondria.

The electrochemical proton gradient that is formed and maintained in the process is used by the ATP synthase (also called complex V) to make ATP (Figure 1). The plant ETC, however, is highly branched. As a consequence, the plant oxidative phosphorylation system has several alternative pathways of electron transport via type II NAD(P)H dehydrogenases and the alternative oxidase (AOX), and additionally, an alternative proton transporter called the uncoupling protein (UCP).

These proteins mediate bypasses around the proton-translocating multi-protein complexes (I–V): type II NAD(P)H dehydrogenases bypass complex I, AOX bypasses complexes III and IV, and UCP bypasses the ATP synthase (Siedow and Umbach 1995; Vanlerberghe and McIntosh 1997; Møller 2001; Vercesi 2001; Millenaar and Lambers 2003; Fernie et al. 2004; Rasmusson et al. 2004).

The type II NAD(P)H dehydrogenases and the AOX transport electrons without pumping protons, whereas the UCP allows proton flow from the intermembrane space to the matrix without ATP synthesis.

The consequence of this is that flux through the energy conservation bypasses will neither contribute to respiratory ATP production, nor will they be controlled by cellular adenylate status. The special features of these energy bypass proteins and discuss the consequences of their existence to the plant.

NAD(P)H Dehydrogenases in Plant Mitochondria

Except for yeast, the ETC of all eukaryotes—including plants—contains a proton-pumping NADH dehydrogenase, complex I, which is inhibited specifically by rotenone. Additionally, plant, fungal and protist mitochondria contain different setups of rotenone-insensitive enzymes capable of oxidizing NADH and/or NAD(P)H and passing the electrons on to ubiquinone.

These NAD(P)H dehydrogenases are classified as type II, whereas complex I belongs to type I. The inner mitochondrial membrane in plants contains such NAD(P)H dehydrogenases, facing either the matrix or the intermembrane space, thus oxidizing NAD(P)H either from the matrix or the cytoplasm.

The outer membrane of all mitochondria contains NADH-ferricyanide and NADH-cytochrome *b* reductase activities that are insensitive to rotenone and to the complex III inhibitor, antimycin A.

There are Three NAD(P)H Dehydrogenases on the Inner Matrix Surface of the Inner Membrane

Much of the information about the number of matrix-facing NAD(P)H dehydrogenases has come from experiments using inside-out inner membrane vesicles. Such vesicles have the matrix surface facing the medium and the external NAD(P)H dehydrogenases hidden inside the vesicles and unable to interfere with the assays.

Recently, this method has been complemented with a technique where one uses a peptide called alamethicin to make a channel for the substrate through the inner membrane. Experiments with these methods show the presence of three NAD(P)H dehydrogenase activities on the matrix side of the inner membrane:

- Complex I pumps protons, is Ca^{2+}-independent and is inhibited by rotenone and diphenylene iodonium (DPI). Complex I has a high affinity for NADH.
- ND_{in}(NADH) does not pump protons, is Ca^{2+}-independent and DPI insensitive. The enzyme has a tenfold lower affinity for NADH than complex I.
- ND_{in}(NADPH) does not pump protons, is Ca^{2+}-dependent and very sensitive to DPI.

Because of the contrasting affinities of complex I and ND_{in}(NADH) for NADH, only complex I is engaged at low matrix concentrations of NADH. However, when the NADH concentration in the matrix increases (e.g., if the ETC is restricted by lack of ADP in the cell), ND_{in}(NADH) can become engaged. When electrons are transported through the type II NAD(P)H dehydrogenases, only the two proton pumping sites of the cytochrome pathway are involved in the flow of electrons from matrix NADH to oxygen. As a consequence, the theoretical ADP/O yield is 1.5 unless the AOX or the UCP are active, in which case it will be even lower.

The External NAD(P)H Dehydrogenases are Ca^{2+}-Dependent

Plant mitochondria can oxidize cytosolic NADH and NAD(P)H directly via external dehydrogenases. The NADH and NAD(P)H dehydrogenase activities are catalyzed by separate enzymes. None of the external dehydrogenases pumps protons, which means that, at most, two sites of proton-pumping are involved when the electrons move from the dehydrogenases to oxygen. So, as for ND_{in}(NADH), the theoretical maximum ADP/O yield is 1.5.

The oxidation of exogenous NADH or NAD(P)H by most plant mitochondria requires Ca^{2+}, and as a result, is inhibited by EDTA and EGTA, chelators of di-and trivalent cations. *In vitro*, the Ca^{2+}concentration needed for half-maximal activity of NADH oxidation is 0.1–0.5 μM, depending on ionic environment and respiratory state This is in the same range as the 0.1–0.2 μM free Ca^{2+} thought to be present in the cytosol of a resting plant cell. Therefore, the external NAD(P)H oxidation may become activated by the increased Ca^{2+}-concentration associated with cellular excitation due to external conditions.

Mammalian mitochondria lack external NAD(P)H dehydrogenases. Instead, they have a Ca^{2+}-dependent glycerol-3-phosphate dehydrogenase on the outer surface of the inner membrane, which donates electrons directly to the ubiquinone pool.

The dihydroxyacetone phosphate formed is converted back into glycerol-3-phosphate by a cytosolic NAD-dependent glycerol-3-phosphate dehydrogenase. In this way, the two low-molecular-weight compounds, glycerol-3-phosphate and dihydroxyacetone phosphate, shuttle redox equivalents from the cytosol, across the outer membrane and to the inner membrane to yield the same amount of ATP as is produced by the plant mechanism. Yeast have both external NADH dehydrogenases and a glycerol-3-phosphate shuttle, and since a mitochondrial glycerol-3-phosphate dehydrogenase has been found in plant mitochondria, plants may also have both systems.

Another way to oxidize cytosolic NAD(P)H that is found in mammalian mitochondria is to use the malate/aspartate shuttle to tranfer reductant (a collective term for high-energy electrons carried by, for example, NAD(P)H, ferredoxin or an organic acid redox couple like malate/oxaloacetate) into the matrix as NADH that can be oxidized by the internal NAD(P)H dehydrogenases. However, it is not clear whether this shuttle can work for import of cytosolic reductant into plant mitochondria.

In the opposite orientation, NADH is shuttled from the matrix to the cytosol via the malate/oxaloacetate shuttle. In this way, the external NADH dehydrogenase can also work as a bypass around complex I. Support for this bypass has been seen in the form of an upregulation of external NADH dehydrogenase activity in mutants where complex I or an internal type II NAD(P)H dehydrogenase has been inactivated.

The Molecular Identities of the Rotenone-Insensitive NAD(P)H Dehydrogenases

The type II NADH dehydrogenases from *E. coli* and yeast are small, single polypeptide proteins of around 50 kDa. They oxidize only NADH and contain FAD as an electron-accepting prosthetic group.

Attempts to purify the type II NAD(P)H dehydrogenases have been unsuccessful in identifying the plant proteins. However, the characterization of potato homologues to the type II NADH dehydrogenases in *E. coli* and yeasts have provided candidate plant proteins to try and assign to the different reported activities.

One protein, NDA1, is internal and the other, NDB1, external of the inner mitochondrial membrane. The latter was shown to be an external calcium-dependent NAD(P)H dehydrogenase. Like a related fungal NAD(P)H dehydrogenase, NDB1 contains a motif for Ca-binding, indicating that in the activation of plant and fungal NAD(P)H dehydrogenase by Ca^{2+}, the ion binds directly to the enzyme.

In Arabidopsis, a weed for which the nuclear genome sequence is known, small families of both NDA-type and NDB-type proteins are encoded by nuclear genes. One of the NDA-type proteins has been shown to be an internal NAD(P)H dehydrogenase. Additionally, a third type of type II NAD(P)H dehydrogenase was found in Arabidopsis, the NDC1. This protein is most closely related to cyanobacterial enzymes, suggesting that the gene entered the eukaryotic domain with the plastid, later to be transferred to the nucleus.

The Alternative Oxidase

The presence of the AOX is one of the features that sets plant (and fungal) mitochondria apart from mammalian mitochondria. This enzyme has therefore received a lot of attention from plant scientists and we now have a fairly good understanding of the regulation of its structure and activity.

The AOX is a quinol-oxygen oxidoreductase and it does not pump protons. The oxygenase transfers electrons from ubiquinol (the reduced form of ubiquinone) to oxygen and generates water as the end-product of the reaction, so four electrons must be transferred to oxygen.

The precise mechanism is not known, but the protein has been shown to contain a catalytical di-iron centre. The functional form of the enzyme is a dimer with the two polypeptides either covalently or non-covalently bound to each other (Figure 2). The predicted structure places the protein embedded into the inner leaflet of the inner mitochondrial membrane, with the active site and the regulatory cysteines exposed to the matrix.

When electrons from NADH oxidation flow through the AOX, at least two of the three sites of proton translocation are bypassed, so energy conservation in the form of ATP is much smaller when the AOX is active. The activity of the AOX must therefore be carefully regulated. There are often several AOX isoforms in each plant species—for example, five genes encoding AOX proteins are present in Arabidopsis. The relative amounts of three isoforms of AOX vary with development, and cellular redox status, as well as external factors.

The AOX is present in the inner membrane as a dimer in which the two subunits can be covalently linked through a disulfide bridge between cysteine residues. The reduced non-covalently linked dimer is the active form, and this form appears to be the predominant one *in vivo*. It was recently shown that a mitochondrial thioredoxin system regulates AOX. Thioredoxins regulate enzymes by reducing disulfide bridges, and it is likely that the regulation involves the redox-active cysteines.

The presence of thioredoxin regulation connects the AOX activation level to the mitochondrial NAD(P)H pool and the oxidation of citric acid cycle metabolites by NADP-utilizing enzymes.

The activity of AOX is also stimulated by 2-oxo organic acids—in particular, pyruvate, which activates by interacting with the same cysteine that is involved in dimerization. This activation may be a feed-forward mechanism that upregulates the capacity of the ETC when the glycolytic end-product is available. Both the cytochrome pathway and the AOX pathway use ubiquinol as the substrate. In the absence of pyruvate, the cytochrome pathway is saturated at 40% Q_{red}/Q_{tot} (40% of the UQ is reduced), at which point the AOX pathway is not engaged at all. The AOX pathway starts accepting electrons at 50% reduction and is not fully saturated at 80%.

In the presence of pyruvate, the AOX becomes engaged at a much lower Q_{red}/Q_{tot} (20%), whereas the cytochrome pathway is unaffected. It is puzzling, however, that the normal *in vivo* concentration of pyruvate appears to be higher than the concentration of pyruvate required to give maximal stimulation of AOX activity *in vitro*. It thus appears that there is little regulation of AOX activity via pyruvate under physiological conditions.

AOX Activity Can Be Measured In Vivo

For many years, benzhydroxamic acid derivatives,, were used to quantify the capacity and activity of the AOX pathway in both isolated mitochondria and intact tissues. AOX capacity was determined from the rate of respiration in the presence of KCN (which blocks the cytochrome pathway), and corrected for residual respiration in the presence of both KCN and SHAM.

AOX *in vivo* activity was determined from the decrease in the rate of respiration when SHAM was added in the absence of KCN. The assumption was that the cytochrome pathway was always used first and that the AOX would be engaged only when the cytochrome pathway was saturated. However, as we have seen above, this assumption is not correct. The consequence of this is that *in vivo* activity of AOX cannot be measured using inhibitors alone.

A better method for measuring AOX *in vivo* activity is now available. It makes use of the fact that cytochrome *c* oxidase and AOX discriminates differently between ^{18}O relative to ^{16}O.

The so-called oxygen discrimination of the sample (tissue or mitochondria) can be measured by mass spectrometry: (1) with KCN to resolve the oxygen discrimination by AOX; (2) with SHAM to resolve the oxygen discrimination by cytochrome *c* oxidase; and (3) without inhibitors to resolve the discrimination value resulting from the activity of both enzymes.

From these values, the percentage contribution by AOX and cytochrome *c* oxidase to the total uninhibited rate of respiration can be calculated. An example of the use of the oxygen discrimination technique with an intact tissue is given below. A remaining problem in the estimation of AOX activity *in vivo* is that measurements can only be made in darkness (to avoid interference from photosynthetic oxygen exchange).

Plant Mitochondria also Contain an Uncoupling Protein

Mammalian mitochondria contain neither type II NAD(P)H dehydrogenases nor AOX. However, they have a UCP that increases the proton permeability of the inner mitochondrial membrane and, in that way, dissipates the proton gradient. This is another mechanism for reducing the ATP production and increasing heat production, for example, for thermoregulation.

Surprisingly, plant mitochondria also contain a UCP. We do not know for certain why plant mitochondria should require two different mechanisms for achieving the same end result. The UCP is related to anion carriers. Its *in vitro* transport activity is dependent on free fatty acids and is inhibited by nucleotides like ATP and GTP but stimulated by superoxide. The latter result indicates that the UCP may protect against oxidative stress.

The Physiological Impact of Energy Bypass Proteins

The respiratory rates of some thermogenic flowers are the highest among plants and, in fact, exceed even those of warm-blooded animals. The external NADH dehydrogenase, AOX and UCP appear particularly active in thermogenic flowers where they are primarily responsible for heat production. At the same time it is clear that the respiratory rate of most plant tissues is insufficient to generate an appreciable increase in tissue temperature.

AOX is up-regulated by cold stress, although this is clearly not to heat the tissue. However, during cold stress, reactive oxygen species (ROS) are formed, and also other stress treatments that lead to elevated ROS will induce AOX. ROS (e.g., superoxide and hydrogen peroxide) are formed as by-products of electron transport under aerobic conditions.

ROS can cause damage to proteins, lipids, and DNA and the cell must therefore limit their formation. AOX helps prevent overreduction of the ETC and thus lowers ROS production. Supporting this, Maxwell et al. (1999) demonstrated the importance of the AOX for limiting ROS formation in living plant cells.

Overexpression of AOX lowered the steady state level of ROS, whereas suppression increased it. This result also points to the mitochondrion as an important site of ROS generation in plant cells. With the recent discovery of ROS stimulation of plant UCP and the increased ROS resistance in plants overexpressing UCP, the picture has become more complex. By decreasing the resistance for electron transport through the proton-pumping complexes, UCP activity may, like AOX, lead to decreased overreduction and ROS formation. Another factor speaking in favour of UCP having a role in ROS protection is that the AOX is inhibited by an ROS product.

Both type II NAD(P)H dehydrogenases and AOX may act as reductant overflow enzymes by reoxidizing surplus NADH to oxygen. This can prevent inhibition of the citric acid cycle under conditions in which the ADP concentration is low, and thus the proton-pumping complexes are restricted by a high membrane potential. In this way, the citric acid cycle can supply carbon skeletons for biosynthesis, even if the ATP level in the cell is high.

The concept of overflow is also relevant for photorespiratory conditions, under which there is a massive flow of glycine that has to be oxidized by the mitochondria. As much as half of the NADH produced in this process is probably exported, but the other half has to be oxidized promptly. There are several reasons to believe that a reductant overflow via energy bypass proteins is involved.

The ND_{in}(NADH) and AOX capacity in leaf mitochondria is higher than in mitochondria from heterotrophic tissue, and both proteins are down-regulated by dark treatment. Consistent with these observations, AOX activity *in vivo* is up-regulated by light (Ribas-Carbo et al. 2000), AOX capacity is induced by greening of etiolated tissue, and expression of the *nda*1 gene [encoding an internal NAD(P)H dehydrogenase] in potato and Arabidopsis is light-dependent.

Mutant tobacco plants, lacking the proton-pumping complex I can still survive, apparently aided by the action of ND_{in}(NADH), which oxidizes matrix NADH with a lower ATP yield.

These plants, however, show decreased photosynthetic rates, indicating that the ND_{in}(NADH) cannot fully compensate for the missing complex I. Also, external NAD(P)H dehydrogenases have been suggested to support photosynthesis. These enzymes may reoxidise NAD(P)H that is shuttled out of the chloroplast in order to balance the NAD(P)H to ATP ratio in the stroma. In summary, there are many indications that mitochondria—especially the energy bypass proteins—have roles in supporting photosynthesis, for example by oxidizing excess NAD(P)H.

The AOX pathway is involved in the acidification/deacidification cycle of CAM plants. CAM plants fix CO_2 in the form of malate during the night. This is done by cytosolic PEP carboxylase, and the malate produced is stored in the vacuole.

The CO_2 fixed is released during the daytime by malic enzymes, located in cytosol and mitochondria. Robinson et al. (1992) used the oxygen discrimination technique to measure the contribution of the AOX pathway to whole-leaf respiration in the CAM-plant, *Kalanchoë daigremontiana.*

They found that the respiratory rate of the leaves increased by 35% in the early hours of the day. This increase was due to increased AOX activity. A more detailed time course showed that the AOX activity increased two-to threefold during the early day at the same time as the malate was metabolized.

As soon as the malate had disappeared by mid-day, the AOX activity returned to the low level observed during the night. A reasonable conclusion is that a substantial amount of the malate from the vacuole is oxidized via malic enzyme in the mitochondria. This generates NADH in the matrix, which is oxidized to a large extent via the AOX. The released CO_2 can then be refixed by the Calvin cycle in the chloroplast.

F_oF_1-ATP Synthases: The Worlds Smallest Rotary Motors

F_oF_1-ATP synthases (also called F-ATPases) are present in the inner membranes of mitochondria, chloroplasts, and bacteria. These large, multisubunit enzymes consist of a water-soluble catalytic complex (F_1) attached to an integral membrane protein complex (F_o) that transports protons across the membrane. The F_1 complex is composed of at least five different types of subunits (three α, three β, one γ, one δ, and one å).

When the F_1 complex is dissociated from the membrane, it is active as an ATPase. In fact, under the appropriate conditions the intact F_oF_1-ATP synthase can run in reverse and act as a proton pump, using the energy of ATP hydrolysis to move H^+ across the membrane. Our understanding of how ATP is synthesized was advanced by Paul Boyer at the University of California, Los Angeles, who proposed the binding-change mechanism for catalysis by F-ATPases. From the detailed understanding of how the ATPases function, he proposed that the binding-change mechanism for ATP synthesis contains three important components:

1. The major energy-requiring step is not the synthesis of ATP from ADP and P_i, but the release of ATP from the enzyme.
2. Substrate is bound and products are released at three separate but interacting catalytic sites, corresponding to the three catalytic subunits (b subunits). Each catalytic site can exist in one of three conformations: tight, loose, or open.
3. The binding changes are coupled to proton transport by rotation of the g subunit. That is, the flow of protons down their electrochemical gradient through the F_0 complex causes the g subunit to rotate.

Rotation of the γ subunit then brings about the conformational changes in the catalytic complex that allow the release of ATP from the enzyme, and the reaction is driven forward. The reverse occurs when the enzyme functions as an ATP-driven proton pump.

The first two predictions of the binding-change model are supported by many lines of evidence, mainly kinetic studies, and are now generally accepted. However, the prediction of a rotary mechanism for coupling proton flow to ATP synthesis has been more difficult to demonstrate. Two major breakthroughs have led to confirmation of the third prediction, as well.

The first breakthrough was the determination of the crystal structure of the F_1-part of bovine mitochondrial ATPase by the laboratory of John Walker in Cambridge, England. The crystal structure showed that the three catalytic β subunits differ in their conformations and in the nucleotide bound to them, consistent with the binding-change mechanism. Even more exciting was the discovery that the γ subunit is inserted like a shaft through the centre of the catalytic complex, which consists of three α subunits and three β subunits arranged alternately in a doughnutlike structure. Moreover, the interface between the γ subunit and the α and β subunits is highly hydrophobic.

The hydrophobicity of the interface minimizes the interactions between the subunits, consistent with the rotation of the γ subunit within the hole formed by the catalytic complex. In other words, the γ subunit looks like a molecular bearing lubricated by a hydrophobic interface. Although many questions were answered by elucidation of the crystal structure of the F_1-ATP synthase, the rotational model cannot be tested by a static "snapshot" of the enzyme. Definitive demonstration of rotation requires a video recording of the spinning of the enzyme in real time. But although they are large for proteins, F_1-ATP synthases are still far too small to be visualized in a light microscope.

To visualize the rotation of the enzyme, Masasuke Yoshida and his colleagues at the Tokyo Institute of Technology came up with an ingenious method to make the enzyme much larger than it is. As Figure below shows, they attached an actin filament labelled with a fluorescent dye to the base of the γ subunit using another protein as a "glue." They then attached the F_1 complex upside down to a glass surface.

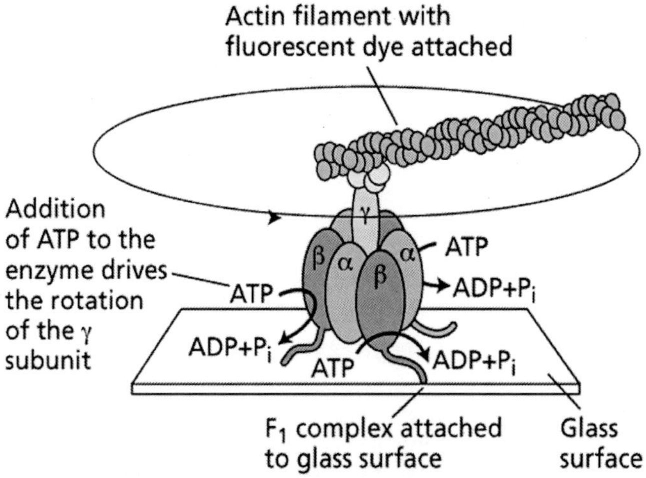

Figure: *A method for visualizing rotation of the g subunit. A fluorescently labelled actin filament was attached to one protruding end of the g subunit. The F_1 complex was then attached upside down to a coverslip. When ATP was added to the coverslip, the actin filament rotated. (After Noji et al. 1997.)*

If the γ subunit rotates with respect to the catalytic complex, the actin filament should swing around with it. Since the filament is very long compared to the ATP synthase (about 1 μm), its rotation should be visible in a fluorescence microscope. In other words, the fluorescently tagged actin filament, which is large enough to visualize in a light microscope, reports the rotation of the γ subunit.

The results were spectacular! When ATP was added to the modified enzyme, the actin filaments were seen to swing around in a circle at as much as 4 revolutions per second in a fluorescence microscope. To give some idea of scale, if you were a γ subunit, this rotation rate would be equivalent to swinging a several hundred meter-long rod around your head at 4 revolutions per second in water! However, the measured velocity is undoubtedly an underestimation of the actual velocity in vivo, because of the enormous torque required to swing such a large mass.

Demonstration of the rotary motion of the γ subunit made it possible to put together a model of how the ATP synthase works. For their contributions to elucidation of the mechanism of ATP synthesis, Paul Boyer and John Walker shared half the Nobel prize in physiology or medicine in 1997. The other half went to Jens Skou for his pioneering work on the K^+,Na^+-ATP synthase, the mammalian counterpart of the plant plasma membrane H^+-ATP synthase.

Transport Into and Out of Plant Mitochondria

Plant mitochondria must exchange metabolites and information (signal transduction) with the cytosol. To give some central examples, the end products of glycolysis, pyruvate and malate, need to be taken up, and citric acid cycle intermediates exported to be used as building blocks in biosynthetic processes.

Likewise, ADP and P_i must be taken up and ATP exported for the mitochondrion to fulfil its role as the powerhouse of the cell. To carry out the transport of this large number of molecules, plant mitochondria possess a family of mitochondrial carrier proteins.

These are transmembrane proteins with a common general structure and they typically transport relatively small charged molecules, like metabolites. In Arabidopsis there are almost 60 members, most of which are not functionally characterized, giving a hint to the complexity of the mitochondrial transport processes. However, mitochondria exchange more than metabolites, and therefore also other types of transporters are involved.

The outer membrane is permeable to molecules of less than 10 kDa, (i.e., all non-polymeric compounds). The permeability results from the presence of pores, which appear to be open in most *in vitro* experiments. Recent studies in animals have characterized pore-forming proteins in the mitochondrial outer membrane, called porins, which appear to regulate transport across the membrane.

Plant mitochondria also have porins but their characterization awaits further research. In our discussion here we will focus on the regulation of mitochondrial transport across the inner mitochondrial membrane. We will first consider the relatively few molecules that can move across the inner membrane without the help of a specific carrier protein. We will then discuss the various metabolite transporters and, finally, briefly look at the transport of proteins and RNA.

Small, Neutral, and Hydrophobic Molecules Diffuse across the Inner Membrane

Irrespective of size, molecules that either are charged or are very hydrophilic can only cross a lipid bilayer with the help of a specific protein carrier.

Small, *neutral* molecules can diffuse across the lipid bilayer and do not require protein transporters. Among those are O_2 and CO_2, two important molecules for respiration. O_2 is needed on the inner surface of the inner membrane where the active sites of both cytochrome *c* oxidase and the alternative oxidase are located.

The CO_2, produced as an end product in the citric acid cycle, needs to diffuse out of the mitochondrion, ultimately to be lost to the surroundings or re-fixed in the chloroplasts. In addition, however, it is possible that CO_2 can also be transported out of the mitochondrion as bicarbonate (HCO_3^-). Mitochondria are osmotically active and swell in a hypotonic medium. This means that the inner membrane is also permeable to H_2O, but it is not clear whether this occurs by diffusion or if aquaporins may be involved.

For technical reasons, it is of importance that several well-known inhibitors of mitochondrial function can diffuse into mitochondria because they are uncharged and hydrophobic. These are rotenone, the specific inhibitor of complex I, antimycin, the specific inhibitor of complex III, HCN, the protonated form of the cyanide (CN^-) inhibitor of complex IV (and other heme-containing enzymes), and oligomycin, the specific inhibitor of the F_oF_1-ATP synthase.

Transport of Inorganic Ions

Inorganic phosphate ($H_2PO_4^-$) is taken up into the matrix in exchange for a hydroxyl ion (HO^-). The energy for the uptake is supplied by the proton gradient (–pH). We will see below that the phosphate gradient, in turn, drives transport of other metabolites.

Several other inorganic ions—Mg^{2+}, Ca^{2+}, K^+, and NH_4^+—are involved in mitochondrial metabolism, but the mechanism by which they pass the inner membrane of plant mitochondria is not well understood. In mammalian mitochondria, uptake of K^+ through a K^+-specific channel, and extrusion via a K^+/H^+ antiporter, is involved in matrix volume regulation and cellular signalling (Garlid and Paucek 2003).

A similar system of K^+ transport is present in plant mitochondria. Both mammalian and plant mitochondria contain an uncoupling protein (UCP). This protein facilitates the movement of protons across the inner membrane and therefore uncouples electron transport and decreases ATP yield.

Transport of Adenylates

One of the important functions of mitochondria is to supply the cell with ATP. To do this, mitochondria have an adenine nucleotide transporter that facilitates the exchange of ADP^{3-} (inward) for ATP^{4-} (outward). This exchange is driven by the membrane potential (inside negative), since there is a net outward movement of one negative charge. In combination with the phosphate carrier reaction, this means that the uptake of one ADP and one phosphate and the export of one ATP (i.e., the transport event connected to mitochondrial ATP synthesis) costs one vectorial proton—the electrical component for adenylate exchange and the “pH component for phosphate uptake.

Transport of Citric Acid Cycle Intermediates and Related Compounds

Pyruvate and malate, the end products of glycolysis, both have transporters in the inner membrane. Pyruvate is taken up in exchange for a hydroxyl ion, and malate^{2-} in exchange for $H_2PO_4^{2-}$.

The transporter for malate also takes succinate^{2-} and is therefore called the dicarboxylate transporter. Malate^{2-} or succinate^{2-} can, in turn, exchange for citrate^{2-} on the tricarboxylate transporter. Malate^{2-} can also exchange for 2-oxoglutarate^{2-} via a special transporter. An interesting transporter, the oxaloacetate transporter, exchanges oxaloacetate^{2-} for malate^{2-} (or citrate^{2-}), which permits the shuttling of reducing equivalents.

The joint activity of these transporters allows plant mitochondria to import or export most citric acid intermediates. Finally, the glutamate/aspartate transporter facilitates the uptake of glutamate. In mammalian mitochondria, glutamate is taken up together with a proton, and it is thus driven by both the electrical and the chemical proton gradient.

A similar transporter is present in plant mitochondria, but a proton co-transport has not been shown, so the transport may be passive in plants. Recently, several additional transporters for metabolites have been identified, and it is likely that certain setups of transporters are only expressed under special developmental conditions, like the mobilization of storage nutrients in developing seedlings.

In several cases the involvement of the mitochondrion in cellular metabolism implies that certain metabolites are taken up across the inner membrane, but the transporter has not yet been described. This is the case for the massive flux of glycine into—and CO_2, serine and NH_4^+ out of—mitochondria in C_3 leaves during photorespiration This is also the case for proline, which accumulates to high concentrations in the cytosol during episodes of plant water stress, and it is then broken down during the rehydration phase by a pathway involving two mitochondrial enzymes.

Transport of Coenzymes

Coenzymes are either synthesized inside the mitochondrion or imported. There is some evidence that, in addition to ATP, at least NAD^+, thiamine pyrophosphate, and coenzyme A are imported across the inner membrane.

The final step in ascorbic acid biosynthesis has recently been shown to be on the outer surface of the inner mitochondrial membrane. In view of the accumulating evidence that ascorbate has an important function in removing ROS inside the mitochondrion, we should expect to find an ascorbate transporter in the inner mitochondrial membrane.

The situation with folate (involved in 1C transfers) is the opposite to that with ascorbate: The only site of synthesis of many of the folate metabolites in the plant cell is the mitochondrial matrix. We can therefore expect that these molecules are exported via specific transporters.

Transport of Reducing Equivalents—Metabolic Shuttles

Mammalian mitochondria cannot oxidize cytosolic NADH or NAD(P)H directly, so they make use of "shuttle mechanisms." An exchange of metabolites by inner membrane transporters takes place through these shuttles so that a reduced compound moves into the matrix in exchange for a more oxidized compound moving out into the cytosol.

In this way, reducing equivalents are transported into the matrix ultimately to be oxidized by the respiratory chain. The shuttles can also be used to transport reducing equivalents in the opposite direction.

Although plant mitochondria can oxidize cytosolic NADH and NAD(P)H directly by the two external NAD(P)H dehydrogenases, they can also use these two metabolic shuttles. The malate/oxaloacetate shuttle uses malate dehydrogenase in the cytosol and in the matrix to catalyze the interconversion of malate (reduced) and oxaloacetate (oxidized).

These two compounds are exchanged on the oxaloacetate transporter. This exchange, however, is not driven by the electrochemical proton gradient across the inner membrane, so it can only move reducing equivalents from a relatively reduced compartment to a relatively oxidized compartment. Since the mitochondrial matrix is more reduced than the cytosol, the oxaloacetate transporter is likely to work primarily in the export of reducing equivalents.

Under photorespiratory conditions, large amounts of glycine are oxidized in the mitochondria to give NADH while, at the same time, NADH is required in the peroxisomes to reduce hydroxypyruvate to glycerate. It is thought that as much as 50% of the matrix NADH is exported, probably via the malate/oxaloacetate shuttle.

The more complex malate/aspartate shuttle involves two matrix enzymes and two cytosolic enzymes as well as two inner membrane transporters. In mammals, it is powered by the electrochemical proton gradient in the direction of the import of reducing equivalents. However, since the glutamate/aspartate exchanger may be passive in plants, this shuttle in plants may similarly be independent of the electrochemical proton gradient. If so, it would (like the malate/oxaloacetate shuttle described above) only be able to function as an exporter of mitochondrial reductant.

Transport of Proteins—Protein Import

Plant mitochondria contain their own DNA (mtDNA) and ribosomes and are capable of synthesizing the proteins encoded in the mtDNA. However, plant mtDNA encodes only a few of the proteins found in the mitochondrion. The great majority of the mitochondrial proteins have to be imported from the cytosol where they are synthesized on free ribosomes. These are synthesized with a presequence or targeting peptide in the N-terminal end, which is recognized by receptors on the outer mitochondrial membrane.

The import of proteins requires a complex machinery with two multisubunit protein complexes—one in the outer membrane (TOM, transporter of the outer membrane) and one in the inner membrane (TIM, transporter of the inner membrane). Concomitant with, or immediately afterwards the import of the protein into the correct mitochondrial compartment, the targeting peptide is often cleaved off to yield the mature and active protein.

Transport of RNA

Mammalian mtDNA encodes the full complement of tRNA required for mitochondrial protein synthesis. In contrast, plant mtDNA only encodes some of the tRNAs required for protein synthesis within the mitochondrion. The missing tRNA must be imported by a yet-to-be-elucidated import mechanism.

The Genetic System in Plant Mitochondria has Some Special Features

As in plastids, the genetic system in mitochondria differs both structurally and functionally from the nuclear genome and its gene expression. On the other hand, mitochondria share several genetic features with bacteria. For example, several genes are transcribed together and the ribosomes are of a prokaryotic type. This is in line with the evolutionary origin of the mitochondrion as an endosymbiont of the α-subclass of Proteobacteria, a group including the intracellular pathogen *Ricketsia prowazekii*.

Plants are also distinctly different from other eukaryotes in their mitochondrial DNA (mtDNA) and its expression. For example, the plant mtDNA is very large and highly variable in size between species. The presence of several repeats enables the genome to recombine, changing the gene order or even causing a split into smaller subgenomic molecules, each carrying a subset of the genes. Though the size of mtDNA varies between plants, the number of genes appears to be similar in all plants.

In the fully sequenced 367 kilobasepair (kb) mtDNA of *Arabidopsis thaliana,* the known genes (constituting 65 kb) are scattered in a background of noncoding sequence. A similar situation can be found in the sequenced mtDNA of sugarbeet.

Plant mtDNA is quite different from the much smaller (16 kb) single mtDNA molecule of mammals in which every base pair has a function in coding for protein or mediating transcription and replication, and neither introns nor repeats exist. The expression of the mitochondrial genes in plants is carried out by RNA polymerases that are unlike the nuclear polymerases, but related to the enzymes in bacteriophages. For several mitochondrial genes, these phage-type polymerases start up transcription from multiple promoters. Many genes in plant mtDNA are broken up into exons by introns. The introns are removed in the processing of the primary RNA into a mature mRNA that can be translated into the correct polypeptide.

The introns in mitochondria belong to the so-called self-splicing type II group, in which splicing is defined by the intron structure. The splicing can also work in the *trans* mode, in which it can join segments from separate RNA molecules, and thus assemble protein-coding reading frames even if the exons are distributed on different primary transcripts. This allows individual exons of a gene to reside independently in the genome, even in different subgenomic DNA molecules. Because mitochondrial genes are often co-transcribed with unrelated genes into a polycistronic primary transcript, RNA processing also includes endonucleolytic reactions, so that each mature mRNA encodes only one protein. Like mRNAs for nuclear encoded genes, mitochondrial mRNAs are polyadenylated. However, instead of stabilizing (the case for nuclear encoded mRNAs), the mitochondrial mRNA polyadenylation mediates degradation. Instead, stem-loops stabilise the mRNA in plant mitochondria.

In plant mitochondria, an additional RNA processing event takes places—RNA editing. In this process, certain cytidines (C) are deaminated into uridines (U). Rarely, U is aminated to C instead.

This means that the coding sequence is changed at the RNA level, in order to restore the correct information for translation where the DNA has diverged. Editing sites mainly occur in DNA encoding protein sequences and functional RNA, and are rare in noncoding mtDNA. In principle, the higher the importance of an individual base, the higher the probability that it constitutes an editing site. Despite the considerable amount of bases that can be edited (up to 15% of the bases of a gene), the functional relevance of this process and the consequent divergence in mtDNA sequence is not understood. Mitochondrial RNA editing has been found in land plants, but not in algae. It also occurs, though to a lesser extent, in plastids. The mitochondrial genome encodes only a small proportion of the mitochondrial proteins; less than 40 as compared to up to thousands of nuclear-encoded mitochondrial proteins. Through evolution, genes have been transferred from the mitochondrion to the nucleus. This is a process that still takes place, at least in plants.

For example, subunit 2 of cytochrome oxidase is encoded by the mitochondrial genome in all investigated eukaryotes except for some legumes and protists. Within the legume family, several intermediates of the gene transfer have been found.

For example, a functional gene can be found in the mitochondrial genome and an unexpressed one in the nuclear genome, or vise versa. In order for a mitochondrial gene to be successfully transferred to the nucleus, the protein-coding frame must be integrated into the chromosome, and several other modifications must be made.

For example: (1) a promoter for nuclear transcription must be inserted at the beginning of the gene, (2) the mRNA should acquire sequences specifying the addition of a poly-A tail, and (3) a targeting sequence is necessary for the cytosolically synthesized protein to be imported into the mitochondrion. For proper function, elements regulating the expression in different cell types and under different conditions must also be included.

Since the coding sequence for most mitochondrial genes in plants is edited at the RNA level, the mitochondrial genomic sequence will not code for a functional protein if inserted into the nucleus.

Therefore, the DNA that is integrated into the nuclear genome must in most cases be made by reverse transcription from mitochondrial mRNA. Recently, upon sequencing the nuclear genome of *Arabidopsis thaliana*, a copy of most of the mtDNA was found integrated in a nuclear chromosome, indicating that large segments of DNA can move between cellular compartments.

However, most of these mitochondrial genes transferred to the nuclear genome cannot produce functional proteins unless bases corresponding to editing sites are corrected by mutation.

However, this is unlikely to take place before other mutations have further disrupted the sequence. As opposed to other mitochondrial genomes, plant mitochondria utilize the universal genetic code. Thus, this factor will not prevent or delay the transfer of genes between plant organelles as it may do in other eukaryotes.

The special features of the mitochondrial genetic system in plants also have consequences for the study of mitochondrial genes. Due to RNA editing and *trans*-splicing, a reading frame is more difficult to identify based only on genomic sequence. Also, protein sequences in databases are automatically translated from genomic DNA sequences without compensating for RNA editing in many cases. This protocol would not work for mitochondrial genes.

Does Respiration Reduce Crop Yields?

Plant respiration can consume an appreciable amount of the carbon fixed each day during photosynthesis over and above the losses due to photorespiration. To what extent can changes in a plants respiratory metabolism affect crop yields?

Attempts to establish a quantitative relationship between respiratory energy metabolism and the various processes going on in the cell have led to a break-up of respiration into two components. In growth respiration, reduced carbon is processed to bring about the addition of new biomass. The other component, maintenance respiration, is needed to keep existing, mature cells in a viable state. Utilization of energy by maintenance respiration is not well understood, but estimates indicate that it can represent more than 50% of the total respiratory flux.

Although numerous questions remain regarding these issues, there are several empirical examples of relations between plant respiration rates and crop yield. In the forage crop, perennial ryegrass (*Lolium perenne*), yield increases of 10% to 20% were correlated with a 20% decrease in the leaf respiration rate.

Similar correlations have been found for other plants, including corn (*Zea mays*) and tall fescue (*Festaca arundinacea*). However, later investigations have shown that "selection for low rates of mature leaf respiration is not an appropriate method to select for high-yielding cultivars in perennial grasses".

Although there is a potential for increasing crop yield through reduction of respiration rates, a better understanding of the sites and mechanisms that control plant respiration is needed before such changes can be exploited commercially in a systematic fashion by plant physiologists, geneticists, and molecular biologists. Furthermore, much remains to be learned about the general applicability of such observations and the conditions under which slower-respiring lines could be at a disadvantage, causing a reduction in crop yields rather than an increase. Two recent studies illustrate the present difficulty of predicting the effect of directed changes at the molecular level on plant productivity:

Nunes-Nesi et al. (2005) found that the productivity of transgenic tomato plants with a reduced activity of mitochondrial malate dehydrogenase was increased as compared to the wild type. Although the respiratory activity of mitochondria isolated from the transgenic plants was unchanged or higher than in the wild type, the rate of leaf respiration was reduced in the transgenic plants. Photosynthesis was markedly increased in the transgenic lines, possibly linked to increased levels of ascorbate.

Sieger et al. (2005) studied the effect of the alternative oxidase on the growth of tobacco cell cultures. Considering the energy-wasteful nature of the alternative oxidase one might expect that a cell line lacking the enzyme would grow faster than the wild type. Secondly, alternative oxidase has a role in the response of plants to oxidative stress, so a cell line lacking the alternative oxidase would also be expected to handle abiotic stress less well.

The results were surprising—the transgenic cell line with a very low expression of alternative oxidase grew as fast as the wild-type cells under normal, nutrient-sufficient conditions, and faster than the wild type under conditions of macronutrient limitation (low phosphate or low nitrogen).

It appears that the alternative oxidase is an important factor in modulating the growth rate in response to nutrient availability. Plant respiration involves an intricate metabolic network of interacting pathways and a complex regulation of gene expression and enzymatic activities. We clearly need to know much more about these interactions before we can predict the effects of changing the expression of single genes on plant productivity.

The Lipid Composition of Membranes Affects the Cell Biology and Physiology of Plants

The characterization of lipid mutants and the experimental modification of lipid composition by molecular-genetic means have now revealed some broad generalizations about the roles of membrane lipids in the cell biology and physiology of plants. However, the biochemical and physical bases for the phenotypes that are observed remain uncertain. Changes in fatty acid composition have been shown to alter chloroplast size and architecture. Mutations that eliminate polyunsaturated fatty acids block photosynthesis and prevent the plant from growing autotrophically.

Surprisingly, such mutant plants can grow on sucrose, indicating that photosynthesis is the only plant function that absolutely requires a polyunsaturated membrane. However, polyunsaturated fatty acids are also required as precursors of signalling compounds which are necessary for such diverse functions as pollen development and plant defence, such as jasmonic acid,.

One of the most extensively studied issues in membrane biology is the relationship between lipid composition and the ability of organisms to adjust to temperature changes. For example, chill-sensitive plants experience sharp reductions in growth rate and development at temperatures between 0 and 12°C. The physical and physiological changes in chill-sensitive plants that are induced by exposure to low temperature, together with the subsequent expression of stress symptoms, are termed chilling injury.

Many economically important crops, such as cotton, soybean, maize, rice, and many tropical and subtropical fruits, are classified as chill sensitive. In contrast, most plants of temperate origin are able to grow and develop at chilling temperatures and are classified as chill-resistant plants.

In attempts to link the biochemical and physiological changes associated with chilling injury with a single "trigger" or site of damage, it has been suggested that the primary event of chilling injury is a transition from a liquid-crystalline phase to a gel phase in the cellular membranes. According to this proposal, the transition from liquid-crystalline phase to gel phase would result in alterations in the metabolism of chilled cells and lead to injury and death of the chill-sensitive plants. The degree of unsaturation of the fatty acids would determine the temperature at which such damage occurs. A similar hypothesis has been proposed for chloroplast membranes in which the levels of disaturated phosphatidylglycerol (molecules containing no *cis* double bonds in the fatty acid chains) would determine the chilling sensitivity of plant species.

However, the most recent research indicates that the relationship of membrane unsaturation to plant temperature responses is more subtle and complex than is suggested in these earlier hypotheses. On the one hand, studies of five different *Arabidopsis* mutants have demonstrated that reduced unsaturation can result in plants that grow well at 22°C but are less robust than wild-type plants when grown at 2 to 5°C. These results were observed even though the lipid changes in most of the mutants are insufficient to cause a lipid phase transition. In addition, the responses of these mutants to low temperature appear quite distinct from classic chilling sensitivity, suggesting that normal chilling injury may not be related to the level of unsaturation of membrane lipids.

In one particular mutant, *fab1*, saturated forms of phosphatidylglycerol account for 43% of the total leaf phosphatidylglycerol—a higher percentage than is found in many chill-sensitive plants. Nevertheless, the mutant was completely unaffected (when compared with wild-type controls) by a range of low-temperature treatments that quickly led to the death of cucumber and other chill-sensitive plants. A complementary series of experiments has been carried out in tobacco, which is a chill-sensitive plant. The transgenic expression of exogenous genes in tobacco has been used specifically to decrease the level of saturated phosphatidylglycerol or to bring about a general increase in membrane unsaturation.

In each case, damage caused by chilling was alleviated to some extent. These new findings make it clear that the extent of membrane unsaturation or the presence of particular lipids, such as disaturated phosphatidylglycerol, can affect the responses of plants to low temperature. However, membrane lipid composition is not the major determinant of chilling sensitivity in plants.

Utilization of Oil Reserves in Cotyledons

Although the pathway for the mobilization of triacylglycerols has been best characterized in castor bean, it seems to be similar in the storage tissues of other oilseeds.

However, not all seeds quantitatively convert fat to sugar. In castor bean, the endosperm degenerates after the fat and protein reserves are fully utilized. In many oilseeds—such as sunflower (*Helianthus annuus*), cotton (*Gossipium hirsutum*), and members of the squash family (Cucurbitaceae)—the cotyledons become green and photosynthesize after the food reserves are used up.

In these tissues only part of the stored lipid is converted to exported carbohydrate. Much of the lipid-derived carbon remains in the cotyledons, where it contributes to the synthesis of chloroplasts and other cellular structures. Acetyl-CoA is also used directly for the production of energy through respiration. As the greening process takes place, there is a transition in the peroxisome population of these cells: Some peroxisomes have fewer of the characteristics of glyoxysomes and more of leaf-type peroxisomes.

Such a transition is in keeping with the decreased requirement for the breakdown of stored lipids and the increased need to metabolize the products of photorespiration as the tissue goes from a heterotrophic to a more autotrophic mode of metabolism.

Assimilation of Mineral Nutrients

Development of a Root Nodule

The establishment of symbiosis between legumes and rhizobia involves the activation of genes in both the host and the symbiont and an elaborated exchange of signals. The formation of a root nodule, the specialized organ from a plant host that contains the symbiotic nitrogen fixing rhizobia involves two simultaneous processes: infection and nodule organogenesis. The different stages in the initiation and development of a root nodule are shown in Figure below. A detailed description of the infection process.

A recent study of the early infection stages and nodule development in two *Medicago* species and the symbiont, *Rhisobium meliloti* showed that extensive reorganization of microtubules occurred early in the infection process in the pericycle and the inner cortex where the nodule primordium forms. The changes in microtubule organization were observed by immunolocalization, and fluorescence and laser confocal microscopy.

These findings show that upon infection, the pericycle is activated first, followed by the activation of the inner cortex, where the nodule primordium forms. These changes are followed by a major reorganization of the microtubular cytoskeleton network in outer tissues, associated with root hair activation and curling, the formation of pre-infection threads, and the initiation and the growth of an infection network.

Measurement of Nitrogen Fixation

Besides N_2, the nitrogenase enzyme can reduce other substrates such as cyanide, azide and acetylene.

Direct measurement of N_2 fixation (reduction) requires a mass spectrometer, instrumentation that is not readily available, but the reduction of acetylene to ethylene can easily be measured by gas chromatography. Acetylene reduction involves two electrons, $C_2H_2 + 2e^- + 2\ H^+ \rightarrow C_2H_4$, as opposed to the eight electrons required for the reduction of N_2 and $2\ H^+$:

$$N_2 + 8e^- + 8H^+ + 16ATP \rightarrow 2NH_3 + H_2 + 16ADP + 16P_i$$

Therefore, reduction of four ethylene molecules correspond to the reduction of one N_2 molecule. The acetylene method has limitations, however: Acetylene blocks gas diffusion into nodules, and rhizosphere microorganisms produce ethylene. For these reasons, the acetylene method is suitable for comparative studies, while mass spectroscopy is used for direct measurements of N_2 reduction, which is required for precise quantification of nitrogen fixation.

The Synthesis of Methionine

Methionine and cysteine are the two sulfur-containing amino acids found in proteins. Methionine is synthesized in plastids from cysteine. In the first step of the pathway, cysteine and O-phosphohomoserine react to form cystathionine via the enzyme cystathionine-γ-synthase. Cystathionine is cleaved into homocysteine, pyruvate, and ammonia by cystathionine-β-lysase. Finally, methionine synthase methylates homocysteine to form methionine (Lea 1997).

4

Plant-water Relations

Water is the most abundant constituent of all physiologically active plant cells. Leaves, for example, have water contents which lie mostly within a range of 55–85% of their fresh weight. Other relatively succulent parts of plants contain approximately the same proportion of water, and even such largely nonliving tissues as wood may be 30–60% water on a fresh-weight basis.

The smallest water contents in living parts of plants occur mostly in dormant structures, such as mature seeds and spores. The great bulk of the water in any plant constitutes a unit system. This water is not in a static condition. Rather it is part of a hydrodynamic system, which in terrestrial plants involves absorption of water from the soil, its translocation throughout the plant, and its loss to the environment, principally in the process known as transpiration.

Cellular Water Relations

The typical mature, vacuolate plant cell constitutes a tiny osmotic system, and this idea is central to any concept of cellular water dynamics. Although the cell walls of most living plant cells are quite freely permeable to water and solutes, the cytoplasmic layer that lines the cell wall is more permeable to some substances than to others.

If a plant cell in a flaccid condition—one in which the cell sap exerts no pressure against the encompassing cytoplasm and cell wall—is immersed in pure water, inward osmosis of water into the cell sap ensues. This gain of water results in the exertion of a turgor pressure against the protoplasm, which in turn is transmitted to the cell wall. This pressure also prevails throughout the mass of solution within the cell. If the cell wall is elastic, some expansion in the volume of the cell occurs as a result of this pressure, although in many kinds of cells this is relatively small.

If a turgid or partially turgid plant cell is immersed in a solution with a greater osmotic pressure than the cell sap, a gradual shrinkage in the volume of the cell ensues; the amount of shrinkage depends upon the kind of cell and its initial degree of turgidity. When the lower limit of cell wall elasticity is reached and there is continued loss of water from the cell sap, the protoplasmic layer begins to recede from the inner surface of the cell wall. Retreat of the protoplasm from the cell wall often continues until it has shrunk toward the centre of the cell, the space between the protoplasm and the cell wall becoming occupied by the bathing solution. This phenomenon is called plasmolysis.

In some kinds of plant cells movement of water occurs principally by the process of imbibition rather than osmosis. The swelling of dry seeds when immersed in water is a familiar example of this process.

Stomatal Mechanism

Various gases diffuse into and out of physiologically active plants. Those gases of greatest physiological significance are carbon dioxide, oxygen, and water vapour. The great bulk of the gaseous exchanges between a plant and its environment occurs through tiny pores in the epidermis that are called stomates. Although stomates occur on many aerial parts of plants, they are most characteristic of, and occur in greatest abundance in, leaves.

Transpiration Process

The term transpiration is used to designate the process whereby water vapour is lost from plants. Although basically an evaporation process, transpiration is complicated by other physical and physiological conditions prevailing in the plant. Whereas loss of water vapour can occur from any part of the plant which is exposed to the atmosphere, the great bulk of all transpiration occurs from the leaves. There are two kinds of foliar transpiration: (1) stomatal transpiration, in which water vapour loss occurs through the stomates, and (2) cuticular transpiration, which occurs directly from the outside surface of epidermal walls through the cuticle. In most species 90% or more of all foliar transpiration is of the stomatal type.

Transpiration is a necessary consequence of the relation of water to the anatomy of the plant, and especially to the anatomy of the leaves. Terrestrial green plants are dependent upon atmospheric carbon dioxide for their survival. In terrestrial vascular plants the principal carbon dioxide–absorbing surfaces are the moist mesophyll cells walls which bound the intercellular spaces in leaves. Ingress of carbon dioxide into these spaces occurs mostly by diffusion through open stomates. When the stomates are open, outward diffusion of water vapour unavoidably occurs, and such stomatal transpiration accounts for most of the water vapour loss from plants. Although transpiration is thus, in effect, an incidental phenomenon, it frequently has marked indirect effects on other physiological processes which occur in the plant because of its effects on the internal water relations of the plant.

Water Translocation

In terrestrial rooted plants practically all of the water which enters a plant is absorbed from the soil by the roots. The water thus absorbed is translocated to all parts of the plant. The mechanism of the "ascent of sap" (all translocated water contains at least traces of solutes) in plants, especially tall trees, was one of the first processes to excite the interest of plant physiologists.

The upward movement of water in plants occurs in the xylem, which, in the larger roots, trunks, and branches of trees and shrubs, is identical with the wood. In the trunks or larger branches of most kinds of trees, however, sap movement is restricted to a few of the outermost annual layers of wood.

Root pressure is generally considered to be one of the mechanisms of upward transport of water in plants. While it is undoubtedly true that root pressure does account for some upward movement of water in certain species of plants at some seasons, various considerations indicate that it can be only a secondary mechanism of water transport.

Upward translocation of water (actually a very dilute sap) is engendered by an increase in the negativity of water potential in the cells of apical organs of plants. Such increases in the negativity of water potentials occur most commonly in the mesophyll cells of leaves as a result of transpiration.

Water Absorption

The successively smaller branches of the root system of any plant terminate ultimately in the root tips, of which there may be thousands and often millions on a single plant. Most absorption of water occurs in the root tip regions, and especially in the root hair zone. Older portions of most roots become covered with cutinized or suberized layers through which only very limited quantities of water can pass.

Whenever the water potential in the peripheral root cells is less than that of the soil water, movement of water from the soil into the root cells occurs. There is some evidence that, under conditions of marked internal water stress, the tension generated in the xylem ducts will be propagated across the root to the peripheral cells. If this occurs, water potentials of greater negativity could develop in peripheral root cells than would otherwise be possible. The absorption mechanism would operate in fundamentally the same way whether or not the water in the root cells passed into a state of tension. The process just described, often called passive absorption, accounts for most of the absorption of water by terrestrial plants.

The phenomenon of root pressure represents another mechanism of the absorption of water. This mechanism is localized in the roots and is often called active absorption. Water absorption of this type only occurs when the rate of transpiration is low and the soil is relatively moist. Although the xylem sap is a relatively dilute solution, its osmotic pressure is usually great enough to engender a more negative water potential than usually exists in the soil water when the soil is relatively moist. A gradient of water potentials can thus be established, increasing in negativity across the epidermis, cortex, and other root tissues, along which the water can move laterally from the soil to the xylem.

Plant Mineral Nutrition

The relationship between plants and all chemical elements other than carbon, hydrogen, and oxygen in the environment. Plants obtain most of their mineral nutrients by extracting them from solution in the soil or the aquatic environment. Mineral nutrients are so called because most have been derived from the weathering of minerals of the Earth's crust. Nitrogen is exceptional in that little occurs in minerals: the primary source is gaseous nitrogen of the atmosphere.

Some of the mineral nutrients are essential for plant growth; others are toxic, and some absorbed by plants may play no role in metabolism. Many are also essential or toxic for the health and growth of animals using plants as food. Six basic facts have been established: (1) plants do not need any of the solid materials in the soil—they cannot even take them up; (2) plants do not need soil microorganisms; (3) plant roots must have a supply of oxygen; (4) all plants require at least 14 mineral nutrients; (5) all of the essential mineral nutrients may be supplied to plants as simple ions of inorganic salts in solution; and (6) all of the essential nutrients must be supplied in adequate but nontoxic quantities. These facts provide a conceptually simple definition of and test for an essential mineral nutrient. A mineral nutrient is regarded as essential if, in its absence, a plant cannot complete its life cycle.

Nutrients which plants require in relatively large amounts, that is, the essential macronutrients, are nitrogen, sulfur, phosphorus, calcium, potassium, and magnesium. Iron is not required in large amounts and hence is regarded as an essential micronutrient or trace element. With the progressive development of better techniques for purifying water and salts, the list of essential nutrients for all plants has expanded to include boron, manganese, zinc, copper, molybdenum, and chlorine. Evidence has accumulated in support of nickel being essential. In addition, sodium and silicon have been shown to be essential for some plants, beneficial to some, and possibly of no benefit to others. Cobalt has also been shown to be essential for the growth of legumes when relying upon atmospheric nitrogen. Claims that two other chemical elements (vanadium and selenium) may be essential micronutrients have still to be firmly established.

Mineral nutrients may be toxic to plants either because the specific nutrient interferes with plant metabolism or because its concentration in combination with others in solution is excessive and interferes with the plant's water relations. Other chemical elements in the environment may also be toxic. High concentrations of salts in soil solutions or aquatic environments may depress their water potential to such an extent that plants cannot obtain sufficient water to germinate or grow. Some desert plants growing in saline soils can accumulate salt concentrations of 20–50% dry weight in their leaves without damage, but salt concentrations of only 1–2% can damage the leaves of many species.

A number of elements interfere directly with other aspects of plant metabolism. Sodium is thought to become toxic when it reaches concentrations in the cytoplasm that depress enzyme activity or damage the structure of organelles, while the toxicity of selenium is probably due to its interference in metabolism of amino acids and proteins. The ions of the heavy metals, cobalt, nickel, chromium, manganese, copper, and zinc are particularly toxic in low concentrations, especially when the concentration of calcium in solution is low; increasing calcium increases the plant's tolerance. Aluminium is toxic only in acid soils. Boron may be toxic in soils over a wide pH range, and is a serious problem for sensitive crops in regions where irrigation waters contain excessive boron or where the soils contain unusually high levels of boron. All plants grow poorly on very acid soils (pH d" 3.5); some plants may grow reasonably well on somewhat less acid soils. Several factors may be involved, and their interactions with plant species are complex. The harmful effects of soil acidity in some areas have been exacerbated by industrial emissions resulting in acid rain and in deposition of substances which increase the acidity on further reaction in the soil, with consequent damage to plants and animals in these ecosystems.

The elemental composition of plants is important to the health and productivity of animals which graze them. With the exception of boron, all elements which are essential for plant growth are also essential for herbivorous mammals. Animals also require sodium, iodine, and selenium and, in the case of ruminant herbivores, cobalt. As a result, animals may suffer deficiencies of any one of this latter group of elements when ingesting plants which are quite healthy but contain low concentrations of these elements. In addition, nutrients in forage may be rendered unavailable to animals through a variety of factors that prevent their absorption from the gut. Plants and animals differ also in their tolerance of high levels of nutrients, sometimes with deleterious results for grazing animals. For example, the toxicity of high concentrations of selenium in plants to animals grazing them, known as selenosis, was recognized when the puzzling and long-known "alkali disease" and "blind staggers" in grazing livestock in parts of the Great Plains of North America were shown to be symptoms of chronic and acute selenium toxicity.

Essential Points in the Nutrition of Plants

The Nature of Plant Growth Processes

The processes concerned in the growth of plants are the subjects of study by plant physiologists and plant biochemists. A comprehensive account of these processes is outside the scope of the present work, the special object of which is to deal with the outward and visible signs of imperfections in the plant's activities caused by faulty mineral nutrition. Nevertheless it is useful to have before us the general features of the main processes involved and to realize that the symptoms we shall be discussing later have a physiological basis, and are not direct and unchangeable signs of the specific deficiencies but result from the derangement of the complicated mechanism of the plant's vital activities.

The main processes involved in plant development may be summarized as follows:

Absorption: Intake of water and mineral elements by the root system.

Carbon assimilation or photosynthesis: Intake of carbon dioxide from the air by the leaves, and reaction of the gas with water in the leaf in the presence of the green chlorophyll to form sugar and free oxygen.

Formation of protoplasm: Protoplasm is the living material of the plant, consisting mainly of proteins, complex compounds of nitrogen built up by the plant from more simple compounds of this element.

Respiration: The combination of oxygen with various food substances synthesized by the plant, especially sugars, whereby energy is produced.

Transpiration: Loss of water from the plant, mainly from the leaves.

Translocation: The movement of materials within the plant.

Storage: Storage of reserve products in various organs and tissues.

During growth there are continuous processes of building up of complex compounds of carbon and nitrogen and breaking down of these into more simple substances, in which water and oxygen are intimately concerned. These processes together comprise plant metabolism.

In the course of the metabolic processes innumerable substances are formed, such as sugars, starch, cellulose, acids, lignin, tannins, amino acids, proteins, amides etc., and many plants also produce special products, as for instance nicotine in the tobacco plant.

For the normal functioning of the above processes there must be an adequate intake of water by the plant to maintain the plant cells in a more or less turgid condition and, since water is being continuously lost at a varying rate from the plant, intake and movement within the plant tissues must be capable of ready adjustment to these changes.

As a result of metabolic activities plants develop special organs of growth and reproduction, each of which has its special characters and makes particular demands on the nutrient supplies of the plant.

With all plants there are well defined seasonal growth cycles. Thus annuals, such as cereals, begin from the seed, give rise to seedlings, which later flower, form grain and ripen off, whilst perennial deciduous trees, such as apples, pears, etc., begin growth in the spring, using stored reserves of food, form leaves, make shoots, blossom and form fruits and subsequently shed their leaves, but meanwhile pass on reserve foods to various storage organs in preparation for the next season's growth.

Coincident with these growth cycles there are well defined chemical cycles of nutrient elements and elaborated products in the leaves, stems and roots, etc. It will be shown later that these cycles are of great importance in considering p-deficiency effects and in diagnosing their causes.

The Plant Environment

Nutritional problems must be considered in relation to all the conditions in which plants live, and not merely in terms of the amounts of plant nutrients contained in or added to the soil. For example, those who are accustomed to growing plants know that the temperature must not be too low or no growth may result or that if too high the plants may be injured.

An optimum temperature is usually recognized and this may vary according as to whether the plant is young or old. Similarly the importance of light is well known and plants may be put in special positions to obtain a maximum supply of light at one stage and may be shaded at another.

The actual duration of the daily period of illumination also affect growth and there are plants which are classified as requiring "long day" conditions to complete their growth cycles and others as needing "short day" conditions. If the special "long" or "short" day periods are not forthcoming for the respective classes of plants requiring these, their growth cycles are abnormal and they may fail entirely to produce flowers, grain or fruit.

The humidity of the atmosphere, as distinct from the water supply in the soil, is of importance in determining the water conditions within the plant, as these are dependent on both water intake by the roots and water loss from the leaves, the latter being largely influenced by the air humidity.

Even the presence of adequate quantities of plant nutrients in the soil is no guarantee that they will be absorbed by the plant roots. It will be shown later how these may be present in forms which are not available to the plants, but even when they would be considered as being present in suitable forms for absorption, other factors may prevent this taking place. An example of this latter condition is afforded in poorly aerated soils where lack of oxygen near the roots may prevent them from actively absorbing mineral nutrients

The problems of such influences in the plant environment as those just mentioned are complicated by the fact that they do not act independently, but their effects are modified by one another. Thus the effects of light intensity or period of daylight may vary with different temperature conditions.

The requirements of plants for different nutrients may be affected by conditions of light, temperature and water supply, and by other factors of the general environment. Thus the need for nitrogen may be less under conditions of relatively low light intensity whereas the need for potash in these circumstances may be greater, these facts being of importance in growing tomatoes under glass. The effect of nitrogen in relation to light may be shown by growing a plant under normal light conditions with insufficient nitrogen, when the leaves will show the well known symptoms of nitrogen deficiency-pale green, yellow, orange and red tints. If such a plant be then shaded, the leaves will turn a darker green and growth may be visibly increased. It can be shown that the lowered light conditions result in an increase of "soluble" a breaking down of proteins, thereby rendering the nitrogen of these available for growth processes.

This interrelationship of environmental factors is well illustrated by an experiment on apple trees at Long Ashton.

Bramley's Seedling trees were grown in compost in large pots and given a small dressing of a nitrogenous fertilizer. Some of the trees were grown in a specially constructed glass house and an equal number in an adjoining wire enclosure. The trees in the enclosure showed severe symptoms of nitrogen deficiency-pale green and yellow leaves, reddish brown barks and highly coloured, red fruits. The condition was corrected by further dressings of nitrogen. In contrast, the trees under glass, where the light was of less intensity and the temperature higher, made vigorous growth, carried large, green leaves and bore large, green fruits.

Iron and zinc deficiency symptoms may be less severe under conditions of low light intensity, whilst boron deficiency effects are less severe and magnesium deficiency effects are more pronounced in wet seasons than in dry ones.

The rate of water absorption is less at lower temperatures than at higher ones and efficient intake is also dependent on good aeration. These facts may result in a water deficit within plants growing in cold, wet soils when the air temperature is high.

The raw materials needed for plant growth consist of carbon dioxide, which is obtained from the atmosphere through the stomata of the leaves, and water and the so-called mineral nutrients, which normally enter the plant through the medium of the roots.

The importance of water and carbon dioxide in the nutrition of plants will be apparent from the facts that water often comprises 80 to 90 % of the total weight of growing plants, and carbon and oxygen together may account for over 80% of their dry matter, i.e., the solid matter remaining after water is removed. As against these large amounts, the mineral nutrients, as measured by the ash content of the plants, i.e., the mineral residue obtained when the organic matter is destroyed by heat, often contribute from 5 to 15% of the dry matter.

It has been shown in recent years that certain organic compounds, known as "growth promoting substances" or "hormones", which occur in plants, and some of which are also present in soils and natural manures, are capable of producing marked growth responses, such as increased root growth, shoot and leaf curvatures, stimulation or suppression of buds, increased fruit setting, prevention of fruit abscission etc.

They appear to perform important functions in the growth of plants. Examples of substances of this kind which can produce growth responses are 13 indole-acetic acid, 13 indole-butyric acid, phenyl acetic acid, A, naphthalene-acetamide, vitamin B_1

It is not at present clear to what extent growth substances are absorbed by plants from soils, although it has been shown that vitamin B_1, which occurs naturally in soils, can be obtained in this way.

The Mineral Nutrients

It has been shown in numerous researches that certain elements are necessary for the healthy growth of plants. They are sometimes spoken of as essential elements and, since some are needed in relatively large quantities and others in very small amounts, the former are referred to as "major" elements and the latter as "minor" or "trace" elements, or as micro nutrients.

The terms "major" and "minor" do not refer to the relative importance of the functions of the elements in plant growth, and for this reason the term "trace" element is preferable for the latter class. Major elements: Nitrogen, phosphorus, calcium, magnesium, potassium, sulfur.

Trace Elements: Iron, manganese, boron, copper, zinc and molybdenum.

(Iron occupies an intermediate position and is usually included in the major elements group. In dealing with field problems it is more convenient to group it with the trace elements.)

In addition, there are, other elements, such as sodium, chlorine and silicon, which produce beneficial effects on the growth of certain plants but which have not so far been shown to be absolutely essential to growth. The element aluminium is of general occurrence in plants, but seems to be without direct nutritional value, although aluminium sulfate is used, because of its acidifying properties, to change the colour of hydrangeas growing on alkaline soils from pink to blue, and aluminium may also exert indirect influences on nutritional processes.

Other elements often occur in plants but they are not known to serve any useful function and frequently they act as plant poisons or toxins.

The nutrient elements can only be absorbed by plants when present in certain forms: nitrogen from nitrates and ammonium salts; phosphorus from phosphates; calcium, magnesium and potassium from their salts; sulfur from sulfates; iron from ferrous or ferric salts (more readily from ferrous salts); manganese from manganous salts; boron from borates; copper and zinc from their salts, and molybdenum from molybdates.

There may appear to be certain exceptions to this statement in practice. For instance, nitrogen may be applied to a soil as "organic" nitrogen, as in hoof meal or urea, and sulfur may be added as the element itself, as in flowers of sulfur, ground sulfur, etc.

In such conditions the added materials are, however, converted into the nitrate and sulfate forms respectively by soil organisms before being absorbed by the plants.

Further points of importance in connection with the absorption of the mineral nutrients by plants are as follows:

(a) They must be absorbed from relatively dilute solutions or the plants will be! injured or even killed.

(b) Certain of the elements slow down the absorption of others into the plant, e.g., calcium slows down potassium and vice versa. The phenomenon is known as "antagonism".

(c) Healthy plants result when the nutrients are absorbed in certain relative proportions. When the proportions are suitable the nutrient medium is said to be "balanced". When ratios between nutrients are too wide, deficiency conditions are created. Thus if a high proportion of nitrogen to potassium is absorbed, the plant will suffer from potassium deficiency.

(d) Nutrients, even though present in the nutrient solution in satisfactory amounts and proportions, may not be absorbed by the plant unless the "reaction" of the solution as regards acidity and alkalinity is satisfactory. The reaction is measured in terms of the pH scale, which is merely a convenient notation for stating the conditions of acidity in the solution (strength or intensity of acidity, not total amount). The neutral point is represented by pH 7.0; below this value the solution is acid and above it is alkaline. Many crop plants prefer a reaction slightly on the acid side-pH 6.0 to 6.5 and extreme values are in the neighborhood of 4.0 on the acid side and 9.0 on the alkaline side.

(e) The nutrient medium must contain an adequate supply of oxygen, i.e., aeration must be satisfactory.

The Functions of The Mineral Nutrients

Knowledge of the main functions of the mineral nutrients is useful in helping us to understand the effects produced by deficiencies of any one of them.

Essential Major Elements

Nitrogen. Nitrogen is a major constituent of several of the most important substances, which occur in plants. It is of outstanding importance among the essential elements in that nitrogen compounds comprise from 40 to 50% of the dry matter of protoplasm, the living substance of plant cells. For this reason nitrogen is required in relatively large quantities in connection with all growth processes in plants. It follows directly from this that without an adequate supply of nitrogen appreciable growth cannot take place and that plants must remain stunted and relatively undeveloped when nitrogen is deficient.

Proteins, which are of great importance in many plant organs, e.g., seeds, are compounds of nitrogen whilst chlorophyll, the green colouring matter of the leaves, also contains the element. From this latter fact it will be apparent that when nitrogen is deficient leaves will contain relatively little chlorophyll, and will thus tend to be pale green in colour. In addition to the above substances, numerous other organic compounds of importance in plants, such as amino acids, amides, and alkaloids, are compounds of nitrogen.

Certain compounds of nitrogen are very mobile in plants, and this enables them readily to mobilize supplies of the element at vital growing points and to transfer stored supplies to points where they are most required. Such transference is common from old tissues to young growing points when supplies of the element are short. This mobility and re-utilization of nitrogen explains why deficiency symptoms of the element always appear first in the older parts of plants and why growing points are the last to be affected.

Phosphorus: This element, like nitrogen, is closely concerned with the vital growth processes in plants as it is a constituent of nucleic acid, and nuclei in which this occurs are essential parts of all living cells. Hence a deficiency of this element will also be expected to result in greatly restricted growth. Phosphorus is also of importance in seeds and in connection with the metabolism of fats. Compounds of phosphorus are concerned with the processes of respiration and with the efficient functioning and utilization of nitrogen. This relationship to nitrogen probably accounts for the fact that several of the symptoms of phosphorus deficiency are identical or similar to those which result from a deficiency of nitrogen, Phosphorus is also of special importance in the processes concerned in root development and the ripening of seeds and fruits.

Calcium: Calcium occurs in plants chiefly in the leaves and the amounts present in seeds and fruits are relatively low. One of its main functions is as a constituent of the cell wall, the middle lamella of which consists largely of calcium pectate. This function appears to be of fundamental importance since, if calcium is replaced by any other of the essential elements, such as magnesium or potassium, the organic materials and mineral salts in the cells are readily leached through the walls. Although a large proportion of the calcium contained in the plant may be soluble in water-as much as 60% in cabbage calcium does not appear to move freely from the older to the younger parts of plants, and hence young tissues contain lower proportions of calcium than older ones. This may explain why calcium deficiency effects begin at the tips of shoots.

Magnesium: The outstanding fact about magnesium is that it is a constituent of chlorophyll, and is essential to the formation of this pigment. As a result, when magnesium is deficient, one of the symptoms commonly shown by plants is chlorosis. Magnesium is also regarded as a carrier of phosphorus in the plant, particularly in connection with the formation of seeds of high oil content, which contain the compound lecithin. The element seems to be very mobile within the plant, and when deficient is apparently transferred from older to younger tissues where it can be re-utilized in the growth processes. This agrees with the observation that signs of magnesium deficiency invariably make their appearance first on the oldest leaves and progress systematically from them towards the youngest ones.

Potassium: Unlike all the other major elements, potassium does not enter into the composition of any of the important plant constituents, such as proteins, chlorophyll, fats and carbohydrates, concerned in plant metabolism. For this reason its role is more difficult to determine, and in spite of much study it cannot be said that the functions of potassium are clearly understood.

The element is present in all parts of plants in large or fairly large proportions. It seems to be of special importance in leaves and at growing points, as these are especially rich in potassium. Probably the whole of the potassium in plants is present in soluble form, and most of it seems to be contained in the cell sap and cytoplasm.

It is outstanding among the nutrient elements for its mobility and solubility within the plant tissues, and these properties no doubt account for the ready way in which potassium can be re-utilized by young tissues when the element is in short supply.

Among the functions which have been attributed to potassium and the processes with which it may be concerned, the following may be mentioned: The formation of carbohydrates and proteins; the regulation of water conditions within the plant cell and of water loss by transpiration; as a catalyst and condensing agent of complex substances; as an accelerator of enzyme action; as contributing to photosynthesis through its radioactive properties. It has been shown in many instances that the potassium content of plants is frequently much higher than is necessary for healthy growth, and it is generally considered that luxury absorption of potassium often takes place.

The great mobility of potassium in plants, its special importance for and its reutilization by young tissues, and its apparent functions as a regulator of plant processes on a large scale are in harmony with the observations that, when potassium is moderately deficient, the effects are seen first in the older tissues and progress from these towards the growing points, but, when the deficiency is acute, growing points are severely affected, and die-back and general collapse of the plants commonly occur.

Sulfur: Sulfur occurs in plants as a constituent of proteins, and of certain volatile compounds such as mustard oil. It seems to be connected with chlorophyll formation, although it is not a constituent of this substance. Its functions in connection with proteins and chlorophyll doubtless account for the similarity of its deficiency effects to those due to deficiency of nitrogen.

Essential Trace Elements

Iron: Iron is closely concerned with chlorophyll formation but is not a constituent of it. Its role appears in this connection to be that of a catalyst. As a result of this function of iron, chlorosis is invariably an outstanding symptom when the element is deficient. Iron may also act as a catalyst, in the role of an oxygen carrier, in respiration.

A point of great importance in connection with iron is its relative immobility in plant tissues. Its mobility seems to be affected by several factors, such as the presence of manganese, potassium deficiency and high light intensity. There is evidence that the amount of chlorophyll is related to "active" iron in plants. It will thus be seen that so-called iron deficiency in the plant may in fact usually mean iron immobility. Lack of mobility may also account for the fact that iron deficiency is first shown in the younger tissues.

Manganese: The functions of manganese are regarded as being closely associated with those of iron and as being concerned with chlorophyll formation. Hence, when manganese is deficient, chlorosis is a common symptom. Manganese may decrease the solubility of iron by oxidation and hence, an abundance of manganese within the plant may lead to iron deficiency and chlorosis. Manganese is regarded as having the functions of a catalyst; its activities being specially concerned with oxidation and reduction reactions within the plant tissues.

Boron: The exact role of boron is not known, but again the evidence as for the other trace elements, suggests that, its functions are those of a catalyst or reaction regulator. It can delay the onset of calcium deficiency effects but cannot replace calcium; it tends to keep calcium soluble; it may act as a regulator of potassium/calcium ratios, and of the absorption of nitrogen; it may be concerned with the oxidation-reduction equilibrium in cells.

Such functions as the above accord with the results which follow from a deficiency of the element, when growth processes show sudden collapse and drastic derangements of metabolism occur.

Zinc and Copper: Although specific functions have not been determined for these elements, here again the evidence points to their roles as catalysts and regulators. Deficiencies of both are associated with chlorosis and a serious general collapse of vital growth processes.

Since catalysts are not used up in the chemical reactions which they promote, we can understand how it comes about that quite small or even minute quantities of the "trace elements", iron, manganese, boron, zinc and copper, may nevertheless be essential to the plant's health and growth.

Elements Sometimes Beneficial

Sodium: As sodium is not strictly an essential element it cannot be expected to have a specific role in the metabolic activities of plants.

Where sodium produces significant effects it is often regarded as a conserver of potassium and as being able partly to replace that element in its role. In no instance, however, has it been shown that sodium can wholly replace potassium where the latter is acutely deficient. In such circumstances, sodium is ineffective as a substitute for potassium, even for sodium-loving plants, such as sugar beet, mangold and barley. Sodium seems to affect the water relations of plants and often enables sugar beet and other crops to withstand drought conditions which would otherwise produce severe adverse effects.

Chlorine: The evidence of the role of chlorine in plants is somewhat contradictory, and no general statement can be made. In tobacco it has been shown to increase the water content of the tissues and to affect carbohydrate metabolism, leading to an accumulation of starch in the leaves. The element is present in plants as chloride, and is wholly soluble.

5

Genetic Material: Properties and Replication

DNA

Deoxyribonucleic acid (DNA) is a nucleic acid that contains the genetic instructions used in the development and functioning of all known living organisms with the exception of some viruses. The main role of DNA molecules is the long-term storage of information. DNA is often compared to a set of blueprints, like a recipe or a code, since it contains the instructions needed to construct other components of cells, such as proteins and RNA molecules. The DNA segments that carry this genetic information are called genes, but other DNA sequences have structural purposes, or are involved in regulating the use of this genetic information.

DNA consists of two long polymers of simple units called nucleotides, with backbones made of sugars and phosphate groups joined by ester bonds. These two strands run in opposite directions to each other and are therefore anti-parallel. Attached to each sugar is one of four types of molecules called bases. It is the sequence of these four bases along the backbone that encodes information. This information is read using the genetic code, which specifies the sequence of the amino acids within proteins. The code is read by copying stretches of DNA into the related nucleic acid RNA, in a process called transcription. Within cells, DNA is organized into long structures called chromosomes. These chromosomes are duplicated before cells divide, in a process called DNA replication. Eukaryotic organisms (animals, plants, fungi, and protists) store most of their DNA inside the cell nucleus and some of their DNA in organelles, such as mitochondria or chloroplasts. In contrast, prokaryotes (bacteria and archaea) store their DNA only in the cytoplasm. Within the chromosomes, chromatin proteins such as histones compact and organize DNA. These compact structures guide the interactions between DNA and other proteins, helping control which parts of the DNA are transcribed.

Properties

DNA is a long polymer made from repeating units called nucleotides. The DNA chain is 22 to 26 Angströms wide (2.2 to 2.6 nanometres), and one nucleotide unit is 3.3 Å (0.33 nm) long. Although each individual repeating unit is very small, DNA polymers can be very large molecules containing millions of nucleotides. For instance, the largest human chromosome, chromosome number 1, is approximately 220 million base pairs long. In living organisms, DNA does not usually exist as a single molecule, but instead as a pair of molecules that are held tightly together. These two long strands entwine like vines, in the shape of a double helix. The nucleotide repeats contain both the segment of the backbone of the molecule, which holds the chain together, and a base, which interacts with the other DNA strand in the helix. A base linked to a sugar is called a nucleoside and a base linked to a sugar and one or more phosphate groups is called a nucleotide. If multiple nucleotides are linked together, as in DNA, this polymer is called a polynucleotide.

The backbone of the DNA strand is made from alternating phosphate and sugar residues. The sugar in DNA is 2-deoxyribose, which is a pentose (five-carbon) sugar. The sugars are joined together by phosphate groups that form phosphodiester bonds between the third and fifth carbon atoms of adjacent sugar rings. These asymmetric bonds mean a strand of DNA has a direction. In a double helix the direction of the nucleotides in one strand is opposite to their direction in the other strand: the strands are *antiparallel*. The asymmetric ends of DNA strands are called the 52 (*five prime*) and 32 (*three prime*) ends, with the 5' end having a terminal phosphate group and the 3' end a terminal hydroxyl group. One major difference between DNA and RNA is the sugar, with the 2-deoxyribose in DNA being replaced by the alternative pentose sugar ribose in RNA.

The DNA double helix is stabilized by hydrogen bonds between the bases attached to the two strands. The four bases found in DNA are adenine (abbreviated A), cytosine (C), guanine (G) and thymine (T). These four bases are attached to the sugar/phosphate to form the complete nucleotide, as shown for adenosine monophosphate.

These bases are classified into two types; adenine and guanine are fused five- and six-membered heterocyclic compounds called purines, while cytosine and thymine are six-membered rings called pyrimidines. A fifth pyrimidine base, called uracil (U), usually takes the place of thymine in RNA and differs from thymine by lacking a methyl group on its ring. Uracil is not usually found in DNA, occurring only as a breakdown product of cytosine. In addition to RNA and DNA, a large number of artificial nucleic acid analogues have also been created to study the proprieties of nucleic acids, or for use in biotechnology.

Grooves

Twin helical strands form the DNA backbone. Another double helix may be found by tracing the spaces, or grooves, between the strands. These voids are adjacent to the base pairs and may provide a binding site. As the strands are not directly opposite each other, the grooves are unequally sized. One groove, the major groove, is 22 wide and the other, the minor groove, is 12 wide. The narrowness of the minor groove means that the edges of the bases are more accessible in the major groove. As a result, proteins like transcription factors that can bind to specific sequences in double-stranded DNA usually make contacts to the sides of the bases exposed in the major groove. This situation varies in unusual conformations of DNA within the cell, but the major and minor grooves are always named to reflect the differences in size that would be seen if the DNA is twisted back into the ordinary B form.

Base Pairing

Each type of base on one strand forms a bond with just one type of base on the other strand. This is called complementary base pairing. Here, purines form hydrogen bonds to pyrimidines, with A bonding only to T, and C bonding only to G. This arrangement of two nucleotides binding together across the double helix is called a base pair.

As hydrogen bonds are not covalent, they can be broken and rejoined relatively easily. The two strands of DNA in a double helix can therefore be pulled apart like a zipper, either by a mechanical force or high temperature. As a result of this complementarity, all the information in the double-stranded sequence of a DNA helix is duplicated on each strand, which is vital in DNA replication. Indeed, this reversible and specific interaction between complementary base pairs is critical for all the functions of DNA in living organisms.

The two types of base pairs form different numbers of hydrogen bonds, AT forming two hydrogen bonds, and GC forming three hydrogen bonds. DNA with high GC-content is more stable than DNA with low GC-content, but contrary to popular belief, this is not due to the extra hydrogen bond of a GC base pair but rather the contribution of stacking interactions (hydrogen bonding merely provides specificity of the pairing, not stability). As a result, it is both the percentage of GC base pairs and the overall length of a DNA double helix that determine the strength of the association between the two strands of DNA.

Long DNA helices with a high GC content have stronger-interacting strands, while short helices with high AT content have weaker-interacting strands. In biology, parts of the DNA double helix that need to separate easily, such as the TATAAT Pribnow box in some promoters, tend to have a high AT content, making the strands easier to pull apart. In the laboratory, the strength of this interaction can be measured by finding the temperature required to break the hydrogen bonds, their melting temperature (also called T_m value). When all the base pairs in a DNA double helix melt, the strands separate and exist in solution as two entirely independent molecules. These single-stranded DNA molecules (*ssDNA*) have no single common shape, but some conformations are more stable than others.

Sense and Antisense

A DNA sequence is called "sense" if its sequence is the same as that of a messenger RNA copy that is translated into protein. The sequence on the opposite strand is called the "antisense" sequence. Both sense and antisense sequences can exist on different parts of the same strand of DNA (i.e. both strands contain both sense and antisense sequences). In both prokaryotes and eukaryotes, antisense RNA sequences are produced, but the functions of these RNAs are not entirely clear. One proposal is that antisense RNAs are involved in regulating gene expression through RNA-RNA base pairing.

A few DNA sequences in prokaryotes and eukaryotes, and more in plasmids and viruses, blur the distinction between sense and antisense strands by having overlapping genes. In these cases, some DNA sequences do double duty, encoding one protein when read along one strand, and a second protein when read in the opposite direction along the other strand. In bacteria, this overlap may be involved in the regulation of gene transcription, while in viruses, overlapping genes increase the amount of information that can be encoded within the small viral genome.

Supercoiling

DNA can be twisted like a rope in a process called DNA supercoiling. With DNA in its "relaxed" state, a strand usually circles the axis of the double helix once every 10.4 base pairs, but if the DNA is twisted the strands become more tightly or more loosely wound. If the DNA is twisted in the direction of the helix, this is positive supercoiling, and the bases are held more tightly together.

If they are twisted in the opposite direction, this is negative supercoiling, and the bases come apart more easily. In nature, most DNA has slight negative supercoiling that is introduced by enzymes called topoisomerases. These enzymes are also needed to relieve the twisting stresses introduced into DNA strands during processes such as transcription and DNA replication.

Alternate DNA Structures

DNA exists in many possible conformations that include A-DNA, B-DNA, and Z-DNA forms, although, only B-DNA and Z-DNA have been directly observed in functional organisms. The conformation that DNA adopts depends on the hydration level, DNA sequence, the amount and direction of supercoiling, chemical modifications of the bases, the type and concentration of metal ions, as well as the presence of polyamines in solution.

The first published reports of A-DNA X-ray diffraction patterns— and also B-DNA used analyses based on Patterson transforms that provided only a limited amount of structural information for oriented fibres of DNA. An alternate analysis was then proposed by Wilkins *et al.*, in 1953, for the *in vivo* B-DNA X-ray diffraction/scattering patterns of highly hydrated DNA fibres in terms of squares of Bessel functions. In the same journal, James D. Watson and Francis Crick presented their molecular modeling analysis of the DNA X-ray diffraction patterns to suggest that the structure was a double-helix.

Although the 'B-DNA form' is most common under the conditions found in cells, it is not a well-defined conformation but a family of related DNA conformations that occur at the high hydration levels present in living cells. Their corresponding X-ray diffraction and scattering patterns are characteristic of molecular paracrystals with a significant degree of disorder. Compared to B-DNA, the A-DNA form is a wider right-handed spiral, with a shallow, wide minor groove and a narrower, deeper major groove. The A form occurs under non-physiological conditions in partially dehydrated samples of DNA, while in the cell it may be produced in hybrid pairings of DNA and RNA strands, as well as in enzyme-DNA complexes. Segments of DNA where the bases have been chemically modified by methylation may undergo a larger change in conformation and adopt the Z form. Here, the strands turn about the helical axis in a left-handed spiral, the opposite of the more common B form. These unusual structures can be recognized by specific Z-DNA binding proteins and may be involved in the regulation of transcription.

Quadruplex Structures

At the ends of the linear chromosomes are specialized regions of DNA called telomeres. The main function of these regions is to allow the cell to replicate chromosome ends using the enzyme telomerase, as the enzymes that normally replicate DNA cannot copy the extreme 32 ends of chromosomes. These specialized chromosome caps also help protect the DNA ends, and stop the DNA repair systems in the cell from treating them as damage to be corrected. In human cells, telomeres are usually lengths of single-stranded DNA containing several thousand repeats of a simple TTAGGG sequence.

These guanine-rich sequences may stabilize chromosome ends by forming structures of stacked sets of four-base units, rather than the usual base pairs found in other DNA molecules. Here, four guanine bases form a flat plate and these flat four-base units then stack on top of each other, to form a stable *G-quadruplex* structure. These structures are stabilized by hydrogen bonding between the edges of the bases and chelation of a metal ion in the centre of each four-base unit. Other structures can also be formed, with the central set of four bases coming from either a single strand folded around the bases, or several different parallel strands, each contributing one base to the central structure. In addition to these stacked structures, telomeres also form large loop structures called telomere loops, or T-loops. Here, the single-stranded DNA curls around in a long circle stabilized by telomere-binding proteins. At the very end of the T-loop, the single-stranded telomere DNA is held onto a region of double-stranded DNA by the telomere strand disrupting the double-helical DNA and base pairing to one of the two strands. This triple-stranded structure is called a displacement loop or D-loop.

Branched DNA

In DNA fraying occurs when non-complementary regions exist at the end of an otherwise complementary double-strand of DNA. However, branched DNA can occur if a third strand of DNA is introduced and contains adjoining regions able to hybridize with the frayed regions of the pre-existing double-strand. Although the simplest example of branched DNA involves only three strands of DNA, complexes involving additional strands and multiple branches are also possible.

Vibration

DNA may carry out low-frequency collective motion as observed by the Raman spectroscopy and analyzed with a quasi-continuum model.

Chemical Modifications

Base Modifications

The expression of genes is influenced by how the DNA is packaged in chromosomes, in a structure called chromatin. Base modifications can be involved in packaging, with regions that have low or no gene expression usually containing high levels of methylation of cytosine bases. For example, cytosine methylation, produces 5-methylcytosine, which is important for X-chromosome inactivation. The average level of methylation varies between organisms - the worm *Caenorhabditis elegans* lacks cytosine methylation, while vertebrates have higher levels, with up to 1% of their DNA containing 5-methylcytosine. Despite the importance of 5-methylcytosine, it can deaminate to leave a thymine base, so methylated cytosines are particularly prone to mutations. Other base modifications include adenine methylation in bacteria, the presence of 5-hydroxymethylcytosine in the brain, and the glycosylation of uracil to produce the "J-base" in kinetoplastids.

Damage

DNA can be damaged by many sorts of mutagens, which change the DNA sequence. Mutagens include oxidizing agents, alkylating agents and also high-energy electromagnetic radiation such as ultraviolet light and X-rays. The type of DNA damage produced depends on the type of mutagen. For example, UV light can damage DNA by producing thymine dimers, which are cross-links between pyrimidine bases. On the other hand, oxidants such as free radicals or hydrogen peroxide produce multiple forms of damage, including base modifications, particularly of guanosine, and double-strand breaks. A typical human cell contains about 150,000 bases that have suffered oxidative damage. Of these oxidative lesions, the most dangerous are double-strand breaks, as these are difficult to repair and can produce point mutations, insertions and deletions from the DNA sequence, as well as chromosomal translocations.

Many mutagens fit into the space between two adjacent base pairs, this is called *intercalation*. Most intercalators are aromatic and planar molecules; examples include ethidium bromide, daunomycin, and doxorubicin. In order for an intercalator to fit between base pairs, the bases must separate, distorting the DNA strands by unwinding of the double helix. This inhibits both transcription and DNA replication, causing toxicity and mutations. As a result, DNA intercalators are often carcinogens, and benzo[*a*]pyrene diol epoxide, acridines, aflatoxin and ethidium bromide are well-known examples. Nevertheless, due to their ability to inhibit DNA transcription and replication, other similar toxins are also used in chemotherapy to inhibit rapidly growing cancer cells.

Biological Functions

DNA usually occurs as linear chromosomes in eukaryotes, and circular chromosomes in prokaryotes. The set of chromosomes in a cell makes up its genome; the human genome has approximately 3 billion base pairs of DNA arranged into 46 chromosomes. The information carried by DNA is held in the sequence of pieces of DNA called genes. Transmission of genetic information in genes is achieved via complementary base pairing.

For example, in transcription, when a cell uses the information in a gene, the DNA sequence is copied into a complementary RNA sequence through the attraction between the DNA and the correct RNA nucleotides. Usually, this RNA copy is then used to make a matching protein sequence in a process called translation, which depends on the same interaction between RNA nucleotides. In alternative fashion, a cell may simply copy its genetic information in a process called DNA replication. The details of these functions are covered in other articles; here we focus on the interactions between DNA and other molecules that mediate the function of the genome.

Genes and Genomes

Genomic DNA is tightly and orderly packed in the process called DNA condensation to fit the small available volumes of the cell. In eukaryotes, DNA is located in the cell nucleus, as well as small amounts in mitochondria and chloroplasts. In prokaryotes, the DNA is held within an irregularly shaped body in the cytoplasm called the nucleoid.

The genetic information in a genome is held within genes, and the complete set of this information in an organism is called its genotype. A gene is a unit of heredity and is a region of DNA that influences a particular characteristic in an organism. Genes contain an open reading frame that can be transcribed, as well as regulatory sequences such as promoters and enhancers, which control the transcription of the open reading frame.

In many species, only a small fraction of the total sequence of the genome encodes protein. For example, only about 1.5% of the human genome consists of protein-coding exons, with over 50% of human DNA consisting of non-coding repetitive sequences.

The reasons for the presence of so much non-coding DNA in eukaryotic genomes and the extraordinary differences in genome size, or *C-value*, among species represent a long-standing puzzle known as the "C-value enigma". However, DNA sequences that do not code protein may still encode functional non-coding RNA molecules, which are involved in the regulation of gene expression.

Some non-coding DNA sequences play structural roles in chromosomes. Telomeres and centromeres typically contain few genes, but are important for the function and stability of chromosomes. An abundant form of non-coding DNA in humans are pseudogenes, which are copies of genes that have been disabled by mutation. These sequences are usually just molecular fossils, although they can occasionally serve as raw genetic material for the creation of new genes through the process of gene duplication and divergence.

Transcription and Translation

A gene is a sequence of DNA that contains genetic information and can influence the phenotype of an organism. Within a gene, the sequence of bases along a DNA strand defines a messenger RNA sequence, which then defines one or more protein sequences. The relationship between the nucleotide sequences of genes and the amino-acid sequences of proteins is determined by the rules of translation, known collectively as the genetic code. The genetic code consists of three-letter 'words' called *codons* formed from a sequence of three nucleotides (e.g. ACT, CAG, TTT). In transcription, the codons of a gene are copied into messenger RNA by RNA polymerase. This RNA copy is then decoded by a ribosome that reads the RNA sequence by base-pairing the messenger RNA to transfer RNA, which carries amino acids. Since there are 4 bases in 3-letter combinations, there are 64 possible codons (4 combinations). These encode the twenty standard amino acids, giving most amino acids more than one possible codon. There are also three 'stop' or 'nonsense' codons signifying the end of the coding region; these are the TAA, TGA and TAG codons.

Replication

Cell division is essential for an organism to grow, but, when a cell divides, it must replicate the DNA in its genome so that the two daughter cells have the same genetic information as their parent. The double-stranded structure of DNA provides a simple mechanism for DNA replication. Here, the two strands are separated and then each strand's complementary DNA sequence is recreated by an enzyme called DNA polymerase. This enzyme makes the complementary strand by finding the correct base through complementary base pairing, and bonding it onto the original strand. As DNA polymerases can only extend a DNA strand in a 52 to 32 direction, different mechanisms are used to copy the antiparallel strands of the double helix. In this way, the base on the old strand dictates which base appears on the new strand, and the cell ends up with a perfect copy of its DNA.

Interactions with Proteins

All the functions of DNA depend on interactions with proteins. These protein interactions can be non-specific, or the protein can bind specifically to a single DNA sequence. Enzymes can also bind to DNA and of these, the polymerases that copy the DNA base sequence in transcription and DNA replication are particularly important.

DNA-binding Proteins

Structural proteins that bind DNA are well-understood examples of non-specific DNA-protein interactions. Within chromosomes, DNA is held in complexes with structural proteins. These proteins organize the DNA into a compact structure called chromatin. In eukaryotes this structure involves DNA binding to a complex of small basic proteins called histones, while in prokaryotes multiple types of proteins are involved. The histones form a disk-shaped complex called a nucleosome, which contains two complete turns of double-stranded DNA wrapped around its surface.

These non-specific interactions are formed through basic residues in the histones making ionic bonds to the acidic sugar-phosphate backbone of the DNA, and are therefore largely independent of the base sequence. Chemical modifications of these basic amino acid residues include methylation, phosphorylation and acetylation. These chemical changes alter the strength of the interaction between the DNA and the histones, making the DNA more or less accessible to transcription factors and changing the rate of transcription. Other non-specific DNA-binding proteins in chromatin include the high-mobility group proteins, which bind to bent or distorted DNA. These proteins are important in bending arrays of nucleosomes and arranging them into the larger structures that make up chromosomes.

A distinct group of DNA-binding proteins are the DNA-binding proteins that specifically bind single-stranded DNA. In humans, replication protein A is the best-understood member of this family and is used in processes where the double helix is separated, including DNA replication, recombination and DNA repair. These binding proteins seem to stabilize single-stranded DNA and protect it from forming stem-loops or being degraded by nucleases.

In contrast, other proteins have evolved to bind to particular DNA sequences. The most intensively studied of these are the various transcription factors, which are proteins that regulate transcription. Each transcription factor binds to one particular set of DNA sequences and activates or inhibits the transcription of genes that have these sequences close to their promoters. The transcription factors do this in two ways. Firstly, they can bind the RNA polymerase responsible for transcription, either directly or through other mediator proteins; this locates the polymerase at the promoter and allows it to begin transcription. Alternatively, transcription factors can bind enzymes that modify the histones at the promoter; this will change the accessibility of the DNA template to the polymerase.

As these DNA targets can occur throughout an organism's genome, changes in the activity of one type of transcription factor can affect thousands of genes. Consequently, these proteins are often the targets of the signal transduction processes that control responses to environmental changes or cellular differentiation and development. The specificity of these transcription factors' interactions with DNA come from the proteins making multiple contacts to the edges of the DNA bases, allowing them to "read" the DNA sequence. Most of these base-interactions are made in the major groove, where the bases are most accessible.

DNA-modifying Enzymes

Nucleases and Ligases

Nucleases are enzymes that cut DNA strands by catalyzing the hydrolysis of the phosphodiester bonds. Nucleases that hydrolyse nucleotides from the ends of DNA strands are called exonucleases, while endonucleases cut within strands. The most frequently used nucleases in molecular biology are the restriction endonucleases, which cut DNA at specific sequences. For instance, the EcoRV enzyme shown to the left recognizes the 6-base sequence 52 -GAT | ATC-32 and makes a cut at the vertical line. In nature, these enzymes protect bacteria against phage infection by digesting the phage DNA when it enters the bacterial cell, acting as part of the restriction modification system. In technology, these sequence-specific nucleases are used in molecular cloning and DNA fingerprinting.

Enzymes called DNA ligases can rejoin cut or broken DNA strands. Ligases are particularly important in lagging strand DNA replication, as they join together the short segments of DNA produced at the replication fork into a complete copy of the DNA template. They are also used in DNA repair and genetic recombination.

Topoisomerases and Helicases

Topoisomerases are enzymes with both nuclease and ligase activity. These proteins change the amount of supercoiling in DNA. Some of these enzymes work by cutting the DNA helix and allowing one section to rotate, thereby reducing its level of supercoiling; the enzyme then seals the DNA break. Other types of these enzymes are capable of cutting one DNA helix and then passing a second strand of DNA through this break, before rejoining the helix. Topoisomerases are required for many processes involving DNA, such as DNA replication and transcription. Helicases are proteins that are a type of molecular motor. They use the chemical energy in nucleoside triphosphates, predominantly ATP, to break hydrogen bonds between bases and unwind the DNA double helix into single strands. These enzymes are essential for most processes where enzymes need to access the DNA bases.

Polymerases

Polymerases are enzymes that synthesize polynucleotide chains from nucleoside triphosphates. The sequence of their products are copies of existing polynucleotide chains - which are called *templates*. These enzymes function by adding nucleotides onto the 32 hydroxyl group of the previous nucleotide in a DNA strand. As a consequence, all polymerases work in a 52 to 32 direction. In the active site of these enzymes, the incoming nucleoside triphosphate base-pairs to the template: this allows polymerases to accurately synthesize the complementary strand of their template. Polymerases are classified according to the type of template that they use.

In DNA replication, a DNA-dependent DNA polymerase makes a copy of a DNA sequence. Accuracy is vital in this process, so many of these polymerases have a proofreading activity. Here, the polymerase recognizes the occasional mistakes in the synthesis reaction by the lack of base pairing between the mismatched nucleotides. If a mismatch is detected, a 32 to 52 exonuclease activity is activated and the incorrect base removed. In most organisms, DNA polymerases function in a large complex called the replisome that contains multiple accessory subunits, such as the DNA clamp or helicases.

RNA-dependent DNA polymerases are a specialized class of polymerases that copy the sequence of an RNA strand into DNA. They include reverse transcriptase, which is a viral enzyme involved in the infection of cells by retroviruses, and telomerase, which is required for the replication of telomeres. Telomerase is an unusual polymerase because it contains its own RNA template as part of its structure.

Transcription is carried out by a DNA-dependent RNA polymerase that copies the sequence of a DNA strand into RNA. To begin transcribing a gene, the RNA polymerase binds to a sequence of DNA called a promoter and separates the DNA strands. It then copies the gene sequence into a messenger RNA transcript until it reaches a region of DNA called the terminator, where it halts and detaches from the DNA. As with human DNA-dependent DNA polymerases, RNA polymerase II, the enzyme that transcribes most of the genes in the human genome, operates as part of a large protein complex with multiple regulatory and accessory subunits.

Genetic Recombination

A DNA helix usually does not interact with other segments of DNA, and in human cells the different chromosomes even occupy separate areas in the nucleus called "chromosome territories". This physical separation of different chromosomes is important for the ability of DNA to function as a stable repository for information, as one of the few times chromosomes interact is during chromosomal crossover when they recombine. Chromosomal crossover is when two DNA helices break, swap a section and then rejoin.

Recombination allows chromosomes to exchange genetic information and produces new combinations of genes, which increases the efficiency of natural selection and can be important in the rapid evolution of new proteins. Genetic recombination can also be involved in DNA repair, particularly in the cell's response to double-strand breaks. The most common form of chromosomal crossover is homologous recombination, where the two chromosomes involved share very similar sequences. Non-homologous recombination can be damaging to cells, as it can produce chromosomal translocations and genetic abnormalities. The recombination reaction is catalyzed by enzymes known as recombinases, such as RAD51. The first step in recombination is a double-stranded break either caused by an endonuclease or damage to the DNA. A series of steps catalyzed in part by the recombinase then leads to joining of the two helices by at least one Holliday junction, in which a segment of a single strand in each helix is annealed to the complementary strand in the other helix. The Holliday junction is a tetrahedral junction structure that can be moved along the pair of chromosomes, swapping one strand for another. The recombination reaction is then halted by cleavage of the junction and re-ligation of the released DNA.

Evolution

DNA contains the genetic information that allows all modern living things to function, grow and reproduce. However, it is unclear how long in the 4-billion-year history of life DNA has performed this function, as it has been proposed that the earliest forms of life may have used RNA as their genetic material. RNA may have acted as the central part of early cell metabolism as it can both transmit genetic information and carry out catalysis as part of ribozymes. This ancient RNA world where nucleic acid would have been used for both catalysis and genetics may have influenced the evolution of the current genetic code based on four nucleotide bases. This would occur, since the number of different bases in such an organism is a trade-off between a small number of bases increasing replication accuracy and a large number of bases increasing the catalytic efficiency of ribozymes. However, there is no direct evidence of ancient genetic systems, as recovery of DNA from most fossils is impossible. This is because DNA will survive in the environment for less than one million years and slowly degrades into short fragments in solution. Claims for older DNA have been made, most notably a report of the isolation of a viable bacterium from a salt crystal 250 million years old, but these claims are controversial.

Uses in Technology

Genetic Engineering

Methods have been developed to purify DNA from organisms, such as phenol-chloroform extraction and manipulate it in the laboratory, such as restriction digests and the polymerase chain reaction. Modern biology and biochemistry make intensive use of these techniques in recombinant DNA technology. Recombinant DNA is a man-made DNA sequence that has been assembled from other DNA sequences. They can be transformed into organisms in the form of plasmids or in the appropriate format, by using a viral vector. The genetically modified organisms produced can be used to produce products such as recombinant proteins, used in medical research, or be grown in agriculture.

Forensics

Forensic scientists can use DNA in blood, semen, skin, saliva or hair found at a crime scene to identify a matching DNA of an individual, such as a perpetrator. This process is formally termed DNA profiling, but may also be called "genetic fingerprinting". In DNA profiling, the lengths of variable sections of repetitive DNA, such as short tandem repeats and minisatellites, are compared between people. This method is usually an extremely reliable technique for identifying a matching DNA. However, identification can be complicated if the scene is contaminated with DNA from several people. DNA profiling was developed in 1984 by British geneticist Sir Alec Jeffreys, and first used in forensic science to convict Colin Pitchfork in the 1988 Enderby murders case.

People convicted of certain types of crimes may be required to provide a sample of DNA for a database. This has helped investigators solve old cases where only a DNA sample was obtained from the scene. DNA profiling can also be used to identify victims of mass casualty incidents. On the other hand, many convicted people have been released from prison on the basis of DNA techniques, which were not available when a crime had originally been committed.

DNA Profiling

DNA profiling (also called DNA testing, DNA typing, or genetic fingerprinting) is a technique employed by forensic scientists to assist in the identification of individuals on the basis of their respective DNA profiles. DNA profiles are encrypted sets of numbers that reflect a person's DNA make-up, which can also be used as the person's identifier. DNA profiling should not be confused with full genome sequencing. It is used in, for example, parental testing and rape investigation.

Although 99.9% of human DNA sequences are the same in every person, enough of the DNA is different to distinguish one individual from another. DNA profiling uses repetitive ("repeat") sequences that are highly variable, called variable number tandem repeats (VNTR). VNTRs loci are very similar between closely related humans, but so variable that unrelated individuals are extremely unlikely to have the same VNTRs. The DNA profiling technique was first reported in 1984 by Sir Alec Jeffreys at the University of Leicester in England, and is now the basis of several national DNA databases. Dr. Jeffreys's genetic fingerprinting was made commercially available in 1987, when a chemical company, ICI, started a blood-testing centre in England.

DNA Profiling Process

The process begins with a sample of an individual's DNA (typically called a "reference sample"). The most desirable method of collecting a reference sample is the use of a buccal swab, as this reduces the possibility of contamination. When this is not available (e.g. because a court order may be needed and not obtainable) other methods may need to be used to collect a sample of blood, saliva, semen, or other appropriate fluid or tissue from personal items (e.g. toothbrush, razor, etc.) or from stored samples (e.g. banked sperm or biopsy tissue). Samples obtained from blood relatives (biological relative) can provide an indication of an individual's profile, as could human remains which had been previously profiled.

A reference sample is then analyzed to create the individual's DNA profile using one of a number of techniques, discussed below. The DNA profile is then compared against another sample to determine whether there is a genetic match.

RFLP Analysis

The first methods for finding out genetics used for DNA profiling involved restriction enzyme digestion, followed by Southern blot analysis. Although polymorphisms can exist in the restriction enzyme cleavage sites, more commonly the enzymes and DNA probes were used to analyze VNTR loci. However, the Southern blot technique is laborious, and requires large amounts of undegraded sample DNA. Also, Karl Brown's original technique looked at many minisatellite loci at the same time, increasing the observed variability, but making it hard to discern individual alleles (and thereby precluding parental testing). These early techniques have been supplanted by PCR-based assays.

PCR Analysis

With the invention of the polymerase chain reaction (PCR) technique, DNA profiling took huge strides forward in both discriminating power and the ability to recover information from very small (or degraded) starting samples. PCR greatly amplifies the amounts of a specific region of DNA, using oligonucleotide primers and a thermostable DNA polymerase.

Early assays such as the HLA-DQ alpha reverse dot blot strips grew to be very popular due to their ease of use, and the speed with which a result could be obtained. However they were not as discriminating as RFLP. It was also difficult to determine a DNA profile for mixed samples, such as a vaginal swab from a sexual assault victim.

Fortunately, the PCR method is readily adaptable for analyzing VNTR loci. In the United States the FBI has standardized a set of 13 VNTR assays for DNA typing, and has organized the CODIS database for forensic identification in criminal cases. Similar assays and databases have been set up in other countries. Also, commercial kits are available that analyze single-nucleotide polymorphisms (SNPs). These kits use PCR to amplify specific regions with known variations and hybridize them to probes anchored on cards, which results in a coloured spot corresponding to the particular sequence variation.

STR Analysis

The method of DNA profiling used today is based on PCR and uses short tandem repeats (STR). This method uses highly polymorphic regions that have short repeated sequences of DNA (the most common is 4 bases repeated, but there are other lengths in use, including 3 and 5 bases). Because unrelated people almost certainly have different numbers of repeat units, STRs can be used to discriminate between unrelated individuals. These STR loci (locations on a chromosome) are targeted with sequence-specific primers and amplified using PCR. The DNA fragments that result are then separated and detected using electrophoresis. There are two common methods of separation and detection, capillary electrophoresis (CE) and gel electrophoresis.

Each STR is polymorphic, however, the number of alleles is small. Typically each STR allele will be shared by around 5 - 20% of individuals. The power of STR analysis comes from looking at multiple STR loci simultaneously. The pattern of alleles can identify an individual quite accurately. Thus STR analysis provides an excellent identification tool. The more STR regions that are tested in an individual the more discriminating the test becomes.

From country to country, different STR-based DNA-profiling systems are in use. In North America, systems which amplify the CODIS 13 core loci are almost universal, while in the UK the SGM+ system (which is compatible with The National DNA Database), is in use. Whichever system is used, many of the STR regions used are the same. These DNA-profiling systems are based on multiplex reactions, whereby many STR regions will be tested at the same time.

The true power of STR analysis is in its statistical power of discrimination. Because the 13 loci that are currently used for discrimination in CODIS are independently assorted (having a certain number of repeats at one locus doesn't change the likelihood of having any number of repeats at any other locus), the product rule for probabilities can be applied. This means that if someone has the DNA type of ABC, where the three loci were independent, we can say that the probability of having that DNA type is the probability of having type A times the probability of having type B times the probability of having type C. This has resulted in the ability to generate match probabilities of 1 in a quintillion (1 with 18 zeros after it) or more.

However, DNA database searches showed much more frequent than expected false DNA matches including one perfect 13 locus match out of only 30,000 DNA samples in Maryland in January 2007.

Moreover, since there are about 12 million monozygotic twins on Earth, that theoretical probability is useless. For example, the actual probability that 2 random people have the same DNA depends on whether there were twins or triplets (etc.) in the family, and the number of loci used in the test. Where twins are common, the probability of matching the DNA is 22 in 1000, or about 2.2 in 100 will have matching DNA.

In practice, the risk of contaminated-matching is much greater than matching a distant relative, such as a sample being contaminated from nearby objects, or from left-over cells transferred from a prior test. Logically, the risk is greater for matching the most common person in the samples: everything collected from, or in contact with, a victim is a major source of contamination for any other samples brought into a lab. For that reason, multiple control-samples are typically tested, to ensure that they stayed clean, when prepared during the same period as the actual test samples. Unexpected matches (or variations) in several control-samples indicates a high probability of contamination for the actual test samples. In a relationship test, the full DNA profiles should differ (except for twins), to prove that a person wasn't actually matched as being related to their own DNA in another sample.

AmpFLP

Another technique, AmpFLP, or amplified fragment length polymorphism was also put into practice during the early 1990s. This technique was also faster than RFLP analysis and used PCR to amplify DNA samples. It relied on variable number tandem repeat (VNTR) polymorphisms to distinguish various alleles, which were separated on a polyacrylamide gel using an allelic ladder (as opposed to a molecular weight ladder). Bands could be visualized by silver staining the gel. One popular locus for fingerprinting was the D1S80 locus.

As with all PCR based methods, highly degraded DNA or very small amounts of DNA may cause allelic dropout (causing a mistake in thinking a heterozygote is a homozygote) or other stochastic effects. In addition, because the analysis is done on a gel, very high number repeats may bunch together at the top of the gel, making it difficult to resolve. AmpFLP analysis can be highly automated, and allows for easy creation of phylogenetic trees based on comparing individual samples of DNA. Due to its relatively low cost and ease of set-up and operation, AmpFLP remains popular in lower income countries.

DNA Family Relationship Analysis

Using PCR technology, DNA analysis is widely applied to determine genetic family relationships such as paternity, maternity, siblingship and other kinships.

During conception, the father's sperm cell and the mother's egg cell, each containing half the amount of DNA found in other body cells, meet and fuse to form a fertilized egg, called a zygote. The zygote contains a complete set of DNA molecules, a unique combination of DNA from both parents. This zygote divides and multiplies into an embryo and later, a full human being.

At each stage of development, all the cells forming the body contain the same DNA—half from the father and half from the mother. This fact allows the relationship testing to use all types of all samples including loose cells from the cheeks collected using buccal swabs, blood or other types of samples.

While a lot of DNA contains information for a certain function, there is some called junk DNA, which is currently used for human identification. At some special locations (called loci) in the junk DNA, predictable inheritance patterns were found to be useful in determining biological relationships. These locations contain specific DNA markers that DNA scientists use to identify individuals. In a routine DNA paternity test, the markers used are Short Tandem Repeats (STRs), short pieces of DNA that occur in highly differential repeat patterns among individuals.

Each person's DNA contains two copies of these markers—one copy inherited from the father and one from the mother. Within a population, the markers at each person's DNA location could differ in length and sometimes sequence, depending on the markers inherited from the parents.

The combination of marker sizes found in each person makes up his/her unique genetic profile. When determining the relationship between two individuals, their genetic profiles are compared to see if they share the same inheritance patterns at a statistically conclusive rate.

For example, the following sample report from this commercial DNA paternity testing laboratory Universal Genetics signifies how relatedness between parents and child is identified on those special markers:

DNA Marker	*Mother*	*Child*	*Alleged father*
D21S11	28, 30	28, 31	29, 31
D7S820	9, 10	10, 11	11, 12
TH01	14, 15	14, 16	15, 16
D13S317	7, 8	7, 9	8, 9
D19S433	14, 16.2	14, 15	15, 17

The partial results indicate that the child and the alleged father's DNA match among these five markers. The complete test results show this correlation on 16 markers between the child and the tested man to draw a conclusion of whether or not the man is the biological father. Scientifically, each marker is assigned with a Paternity Index (PI), which is a statistical measure of how powerfully a match at a particular marker indicates paternity. The PI of each marker is multiplied with each other to generate the Combined Paternity Index (CPI), which indicates the overall probability of an individual being the biological father of the tested child relative to any random man from the entire population of the same race. The CPI is then converted into a Probability of Paternity showing the degree of relatedness between the alleged father and child.

The DNA test report in other family relationship tests, such as grandparentage and siblingship tests, is similar to a paternity test report. Instead of the Combined Paternity Index, a different value, such as a Siblingship Index, is reported. The report shows the genetic profiles of each tested person. If there are markers shared among the tested individuals, the probability of biological relationship is calculated to determine how likely the tested individuals share the same markers due to a blood relationship.

Y-chromosome Analysis

Recent innovations have included the creation of primers targeting polymorphic regions on the Y-chromosome (Y-STR), which allows resolution of a mixed DNA sample from a male and female and/or cases in which a differential extraction is not possible. Y-chromosomes are paternally inherited, so Y-STR analysis can help in the identification of paternally related males. Y-STR analysis was performed in the Sally Hemings controversy to determine if Thomas Jefferson had sired a son with one of his slaves.

Mitochondrial Analysis

For highly degraded samples, it is sometimes impossible to get a complete profile of the 13 CODIS STRs. In these situations, mitochondrial DNA (mtDNA) is sometimes typed due to there being many copies of mtDNA in a cell, while there may only be 1-2 copies of the nuclear DNA.

Forensic scientists amplify the HV1 and HV2 regions of the mtDNA, then sequence each region and compare single-nucleotide differences to a reference. Because mtDNA is maternally inherited, directly linked maternal relatives can be used as match references, such as one's maternal grandmother's daughter's son. A difference of two or more nucleotides is generally considered to be an exclusion. Heteroplasmy and poly-C differences may throw off straight sequence comparisons, so some expertise on the part of the analyst is required. mtDNA is useful in determining clear identities, such as those of missing people when a maternally linked relative can be found. mtDNA testing was used in determining that Anna Anderson was not the Russian princess she had claimed to be, Anastasia Romanov. mtDNA can be obtained from such material as hair shafts and old bones/teeth..

DNA Databases

There are now several DNA databases in existence around the world. Some are private, but most of the largest databases are government controlled. The United States maintains the largest DNA database, with the Combined DNA Index System, holding over 5 million records as of 2007. The United Kingdom maintains the National DNA Database (NDNAD), which is of similar size, despite the UK's smaller population. The size of this database, and its rate of growth, is giving concern to civil liberties groups in the UK, where police have wide-ranging powers to take samples and retain them even in the event of acquittal. The U.S. Patriot Act of the United States provides a means for the U.S. government to get DNA samples from other countries if they are either a division of, or head office of, a company operating in the U.S. Under the act, the American offices of the company can't divulge to their subsidiaries/offices in other countries the reasons that these DNA samples are sought or by whom.

When a match is made from a National DNA Databank to link a crime scene to an offender who has provided a DNA Sample to a databank that link is often referred to as a *cold hit*. A cold hit is of value in referring the police agency to a specific suspect but is of less evidential value than a DNA match made from outside the DNA Databank.

Considerations when Evaluating DNA Evidence

In the early days of the use of genetic fingerprinting as criminal evidence, juries were often swayed by spurious statistical arguments by defence lawyers along these lines: given a match that had a 1 in 5 million probability of occurring by chance, the lawyer would argue that this meant that in a country of say 60 million people there were 12 people who would also match the profile. This was then translated to a 1 in 12 chance of the suspect being the guilty one.

This argument is not sound unless the suspect was drawn at random from the population of the country. In fact, a jury should consider how likely it is that an individual matching the genetic profile would also have been a suspect in the case for other reasons. Another spurious statistical argument is based on the false assumption that a 1 in 5 million probability of a match automatically translates into a 1 in 5 million probability of innocence and is known as the prosecutor's fallacy.

When using RFLP, the theoretical risk of a coincidental match is 1 in 100 billion (100,000,000,000), although the practical risk is actually 1 in 1000 because monozygotic twins are 0.2% of the human population. Moreover, the rate of laboratory error is almost certainly higher than this, and often actual laboratory procedures do not reflect the theory under which the coincidence probabilities were computed. For example, the coincidence probabilities may be calculated based on the probabilities that markers in two samples have bands in *precisely* the same location, but a laboratory worker may conclude that similar—but not precisely identical—band patterns result from identical genetic samples with some imperfection in the agarose gel. However, in this case, the laboratory worker increases the coincidence risk by expanding the criteria for declaring a match.

Recent studies have quoted relatively high error rates which may be cause for concern. In the early days of genetic fingerprinting, the necessary population data to accurately compute a match probability was sometimes unavailable. Between 1992 and 1996, arbitrary low ceilings were controversially put on match probabilities used in RFLP analysis rather than the higher theoretically computed ones. Today, RFLP has become widely disused due to the advent of more discriminating, sensitive and easier technologies.

STRs do not suffer from such subjectivity and provide similar power of discrimination (1 in 10^13 for unrelated individuals if using a full SGM+ profile) It should be noted that figures of this magnitude are not considered to be statistically supportable by scientists in the UK, for unrelated individuals with full matching DNA profiles a match probability of 1 in a billion is considered statistically supportable (Since 1998 the DNA profiling system supported by The National DNA Database in the UK is the SGM+ DNA profiling system which includes 10 STR regions and a sex indicating test. However, with any DNA technique, the cautious juror should not convict on genetic fingerprint evidence alone if other factors raise doubt. Contamination with other evidence (secondary transfer) is a key source of incorrect DNA profiles and raising doubts as to whether a sample has been adulterated is a favorite defence technique. More rarely, chimerism is one such instance where the lack of a genetic match may unfairly exclude a suspect.

Evidence of Genetic Relationship

It's also possible to use DNA profiling as evidence of genetic relationship, but testing that shows no relationship isn't absolutely certain. While almost all individuals have a single and distinct set of genes, rare individuals, known as "chimeras", have at least two different sets of genes. There have been several cases of DNA profiling that falsely "proved" that a mother was unrelated to her children.

Fake DNA Evidence

The value of DNA evidence has to be seen in light of recent cases where criminals planted fake DNA samples at crime scenes. In one case, a criminal even planted fake DNA evidence in his own body: Dr. John Schneeberger raped one of his sedated patients in 1992 and left semen on her underwear. Police drew what they believed to be Schneeberger's blood and compared its DNA against the crime scene semen DNA on three occasions, never showing a match. It turned out that he had surgically inserted a Penrose drain into his arm and filled it with foreign blood and anticoagulants.

In a study conducted by the life science company Nucleix and published in the journal Forensic Science International, scientists found that an In vitro synthesized sample of DNA matching any desired genetic profile can be constructed using standard molecular biology techniques without obtaining any actual tissue from that person.

DNA Evidence as Evidence in Criminal Trials

Familial Searching

Familial searching is the use of family members' DNA to identify a closely related suspect in jurisdictions where large DNA databases exist, but no exact match has been found. The first successful use of the practice was in a UK case where a man was convicted of manslaughter when he threw a brick stained with his own blood into a moving car.

Police could not get an exact match to the UK's DNA database because the man had no criminal convictions, but police implicated him using a close relative's DNA. The technique was used to catch a Los Angeles serial killer known as the "Grim Sleeper" in 2010. However, critics have challenged the technology as "deeply antithetical to American values" and likely to lead to excess investigation of racial and ethnic minorities.

Surreptitious DNA Collecting

Police forces may collect DNA samples without the suspects' knowledge, and use it as evidence. Legality of this mode of proceeding has been questioned in Australia. In the United States, it has been accepted, courts often claiming that there was no expectation of privacy, citing California v. Greenwood (1985), during which the Supreme Court held that the Fourth Amendment does not prohibit the warrantless search and seizure of garbage left for collection outside the curtilage of a home. Critics of this practice underline the fact that this analogy ignores that "most people have no idea that they risk surrendering their genetic identity to the police by, for instance, failing to destroy a used coffee cup.

Moreover, even if they do realize it, there is no way to avoid abandoning one's DNA in public." In the UK, the Human Tissue Act 2004 prohibited private individuals from covertly collecting biological samples (hair, fingernails, etc.) for DNA analysis, but excluded medical and criminal investigations from the offense.

England and Wales

Evidence from an expert who has compared DNA samples must be accompanied by evidence as to the sources of the samples and the procedures for obtaining the DNA profiles. The judge must ensure that the jury must understand the significance of DNA matches and mismatches in the profiles. The judge must also ensure that the jury does not confuse the 'match probability' (the probability that a person that is chosen at random has a matching DNA profile to the sample from the scene) with the 'likelihood ratio' (the probability that a person with matching DNA committed the crime). In *R v. Doheny*, EWCA Crim 728. Phillips LJ gave this example of a summing up, which should be carefully tailored to the particular facts in each case:

Members of the Jury, if you accept the scientific evidence called by the Crown, this indicates that there are probably only four or five white males in the United Kingdom from whom that semen stain could have come. The Defendant is one of them. If that is the position, the decision you have to reach, on all the evidence, is whether you are sure that it was the Defendant who left that stain or whether it is possible that it was one of that other small group of men who share the same DNA characteristics.

Juries should weigh up conflicting and corroborative evidence, using their own common sense and not by using mathematical formulae, such as Bayes' theorem, so as to avoid "confusion, misunderstanding and misjudgment".

Presentation and Evaluation of Evidence of Partial or Incomplete DNA Profiles

R v Bates (2006) EWCA Crim 1395 Moore-Bick LJ said: "We can see no reason why partial profile DNA evidence should not be admissible provided that the jury are made aware of its inherent limitations and are given a sufficient explanation to enable them to evaluate it.

There may be cases where the match probability in relation to all the samples tested is so great that the judge would consider its probative value to be minimal and decide to exclude the evidence in the exercise of his discretion, but this gives rise to no new question of principle and can be left for decision on a case by case basis.

However, the fact that there exists in the case of all partial profile evidence the possibility that a "missing" allele might exculpate the accused altogether does not provide sufficient grounds for rejecting such evidence. In many there is a possibility (at least in theory) that evidence exists which would assist the accused and perhaps even exculpate him altogether, but that does not provide grounds for excluding relevant evidence that is available and otherwise admissible, though it does make it important to ensure that the jury are given sufficient information to enable them to evaluate that evidence properly".

DNA Testing in the US

There are state laws on DNA profiling in all 50 states of the United States. Detailed information on database laws in each state can be found at the National Conference of State Legislatures website.

Development of Artificial DNA

In August 2009, scientists in Israel stunned the forensic sciences and raised serious questions concerning the use of DNA by law enforcement as the ultimate method of identification. In a paper published in the journal *Forensic Science International: Genetics*, the Israeli researchers demonstrated that it is possible to manufacture DNA in a laboratory, and thus falsify DNA evidence. The scientists fabricated saliva and blood samples, which originally contained DNA from a person other than the ostensible donor of the blood and saliva.

The researchers also showed that, using a DNA database, it is possible to take information from a profile and manufacture DNA to match it, and that this can be done without access to any actual DNA from the person whose DNA they are duplicating. The synthetic DNA oligos required for the procedure are common in molecular laboratories.

Dr. Daniel Frumkin, lead author on the paper, was quoted in *The New York Times* as saying, "You can just engineer a crime scene... any biology undergraduate could perform this."

Dr. Frumkin perfected a test that can forensically differentiate real DNA samples from fake ones. His test detects epigenetic modifications, in particular, DNA methylation. Seventy percent of the DNA in any human genome is methylated, meaning it contains methyl group modifications within a CpG dinucleotide context. Methylation at the promoter region is associated with gene silencing. The synthetic DNA lacks this epigenetic modification, which allows the test to distinguish manufactured DNA from original, genuine, DNA.

It is unknown how many, if any, police departments currently use the test, which appears to be a serious issue in light of Frumkin's claim that the DNA manufacturing procedure is within the grasp of any undergraduate biology student. No police lab has publicly announced that it is using the new test to verify DNA results, while FSI Genetics says that any forensic laboratory doing DNA identification should adopt this test to authenticate its results as "real" DNA.

Bioinformatics

Bioinformatics involves the manipulation, searching, and data mining of DNA sequence data. The development of techniques to store and search DNA sequences have led to widely applied advances in computer science, especially string searching algorithms, machine learning and database theory. String searching or matching algorithms, which find an occurrence of a sequence of letters inside a larger sequence of letters, were developed to search for specific sequences of nucleotides. In other applications such as text editors, even simple algorithms for this problem usually suffice, but DNA sequences cause these algorithms to exhibit near-worst-case behaviour due to their small number of distinct characters.

The related problem of sequence alignment aims to identify homologous sequences and locate the specific mutations that make them distinct. These techniques, especially multiple sequence alignment, are used in studying phylogenetic relationships and protein function.

Data sets representing entire genomes' worth of DNA sequences, such as those produced by the Human Genome Project, are difficult to use without annotations, which label the locations of genes and regulatory elements on each chromosome. Regions of DNA sequence that have the characteristic patterns associated with protein- or RNA-coding genes can be identified by gene finding algorithms, which allow researchers to predict the presence of particular gene products in an organism even before they have been isolated experimentally.

DNA Nanotechnology

DNA nanotechnology uses the unique molecular recognition properties of DNA and other nucleic acids to create self-assembling branched DNA complexes with useful properties. DNA is thus used as a structural material rather than as a carrier of biological information.

This has led to the creation of two-dimensional periodic lattices (both tile-based as well as using the "DNA origami" method) as well as three-dimensional structures in the shapes of polyhedra. Nanomechanical devices and algorithmic self-assembly have also been demonstrated, and these DNA structures have been used to template the arrangement of other molecules such as gold nanoparticles and streptavidin proteins.

History and Anthropology

Because DNA collects mutations over time, which are then inherited, it contains historical information, and, by comparing DNA sequences, geneticists can infer the evolutionary history of organisms, their phylogeny. This field of phylogenetics is a powerful tool in evolutionary biology. If DNA sequences within a species are compared, population geneticists can learn the history of particular populations. This can be used in studies ranging from ecological genetics to anthropology; For example, DNA evidence is being used to try to identify the Ten Lost Tribes of Israel.

DNA has also been used to look at modern family relationships, such as establishing family relationships between the descendants of Sally Hemings and Thomas Jefferson. This usage is closely related to the use of DNA in criminal investigations detailed above. Indeed, some criminal investigations have been solved when DNA from crime scenes has matched relatives of the guilty individual.

History of DNA Research

DNA was first isolated by the Swiss physician Friedrich Miescher who, in 1869, discovered a microscopic substance in the pus of discarded surgical bandages. As it resided in the nuclei of cells, he called it "nuclein". In 1919, Phoebus Levene identified the base, sugar and phosphate nucleotide unit. Levene suggested that DNA consisted of a string of nucleotide units linked together through the phosphate groups. However, Levene thought the chain was short and the bases repeated in a fixed order. In 1937 William Astbury produced the first X-ray diffraction patterns that showed that DNA had a regular structure.

In 1928, Frederick Griffith discovered that traits of the "smooth" form of the *Pneumococcus* could be transferred to the "rough" form of the same bacteria by mixing killed "smooth" bacteria with the live "rough" form. This system provided the first clear suggestion that DNA carries genetic information—the Avery-MacLeod-McCarty experiment—when Oswald Avery, along with coworkers Colin MacLeod and Maclyn McCarty, identified DNA as the transforming principle in 1943. DNA's role in heredity was confirmed in 1952, when Alfred Hershey and Martha Chase in the Hershey-Chase experiment showed that DNA is the genetic material of the T2 phage.

In 1953, James D. Watson and Francis Crick suggested what is now accepted as the first correct double-helix model of DNA structure in the journal *Nature*. Their double-helix, molecular model of DNA was then based on a single X-ray diffraction image (labeled as "Photo 51") taken by Rosalind Franklin and Raymond Gosling in May 1952, as well as the information that the DNA bases are paired — also obtained through private communications from Erwin Chargaff in the previous years. Chargaff's rules played a very important role in establishing double-helix configurations for B-DNA as well as A-DNA.

Experimental evidence supporting the Watson and Crick model were published in a series of five articles in the same issue of *Nature*. Of these, Franklin and Gosling's paper was the first publication of their own X-ray diffraction data and original analysis method that partially supported the Watson and Crick model; this issue also contained an article on DNA structure by Maurice Wilkins and two of his colleagues, whose analysis and *in vivo* B-DNA X-ray patterns also supported the presence *in vivo* of the double-helical DNA configurations as proposed by Crick and Watson for their double-helix molecular model of DNA in the previous two pages of *Nature*.

In 1962, after Franklin's death, Watson, Crick, and Wilkins jointly received the Nobel Prize in Physiology or Medicine. However, Nobel rules of the time allowed only living recipients, but a vigorous debate continues on who should receive credit for the discovery.

In an influential presentation in 1957, Crick laid out the "Central Dogma" of molecular biology, which foretold the relationship between DNA, RNA, and proteins, and articulated the "adaptor hypothesis".

Final confirmation of the replication mechanism that was implied by the double-helical structure followed in 1958 through the Meselson-Stahl experiment. Further work by Crick and coworkers showed that the genetic code was based on non-overlapping triplets of bases, called codons, allowing Har Gobind Khorana, Robert W. Holley and Marshall Warren Nirenberg to decipher the genetic code. These findings represent the birth of molecular biology.

Protein Synthesis

Protein synthesis is the process in which cells build proteins. The term is sometimes used to refer only to protein translation but more often it refers to a multi-step process, beginning with amino acid synthesis and transcription of nuclear DNA into messenger RNA, which is then used as input to translation.

The cistron DNA is transcribed into a variety of RNA intermediates. The last version is used as a template in synthesis of a polypeptide chain. Proteins can often be synthesized directly from genes by translating mRNA. When a protein needs to be available on short notice or in large quantities, a protein precursor is produced. A proprotein is an inactive protein containing one or more inhibitory peptides that can be activated when the inhibitory sequence is removed by proteolysis during posttranslational modification.

A preprotein is a form that contains a signal sequence (an N-terminal signal peptide) that specifies its insertion into or through membranes; i.e., targets them for secretion. The signal peptide is cleaved off in the endoplasmic reticulum. Preproproteins have both sequences (inhibitory and signal) still present.

For synthesis of protein, a succession of tRNA molecules charged with appropriate amino acids have to be brought together with an mRNA molecule and matched up by base-pairing through their anti-codons with each of its successive codons. The amino acids then have to be linked together to extend the growing protein chain, and the tRNAs, relieved of their burdens, have to be released. This whole complex of processes is carried out by a giant multimolecular machine, the ribosome, formed of two main chains of RNA, called ribosomal RNA (rRNA), and more than 50 different proteins.

This molecular juggernaut latches onto the end of an mRNA molecule and then trundles along it, capturing loaded tRNA molecules and stitching together the amino acids they carry to form a new protein chain. Protein biosynthesis, although very similar, is different for prokaryotes and eukaryotes.

Amino Acid Synthesis

Amino acid synthesis is the set of biochemical processes (metabolic pathways) by which the various amino acids are produced from other compounds. The substrates for these processes are various compounds in the organism's diet or growth media. Not all organisms are able to synthesise all amino acids, for example humans are only able to synthesise 12 of the 20 standard amino acids.

A fundamental problem for biological systems is to obtain nitrogen in an easily usable form. This problem is solved by certain microorganisms capable of reducing the inert Na"N molecule (nitrogen gas) to two molecules of ammonia in one of the most remarkable reactions in biochemistry.

Nitrogen in the form of ammonia is the source of nitrogen for all the amino acids. The carbon backbones come from the glycolytic pathway, the pentose phosphate pathway, or the citric acid cycle.

In amino acid production, one encounters an important problem in biosynthesis— namely, stereochemical control. Because all amino acids except glycine are chiral, biosynthetic pathways must generate the correct isomer with high fidelity.

In each of the 19 pathways for the generation of chiral amino acids, the stereochemistry at the á-carbon atom is established by a transamination reaction that involves pyridoxal phosphate. Almost all the transaminases that catalyze these reactions descend from a common ancestor, illustrating once again that effective solutions to biochemical problems are retained throughout evolution.

Biosynthetic pathways are often highly regulated such that building blocks are synthesized only when supplies are low. Very often, a high concentration of the final product of a pathway inhibits the activity of enzymes that function early in the pathway. Often present are allosteric enzymes capable of sensing and responding to concentrations of regulatory species. These enzymes are similar in functional properties to aspartate transcarbamoylase and its regulators. Feedback and allosteric mechanisms ensure that all twenty amino acids are maintained in sufficient amounts for protein synthesis and other processes.

Amino acids are synthesized from α-ketoacids, and later transaminated from another aminoacid, usually Glutamate. The enzyme involved in this reaction is an aminotransferase.

Glutamate itself is formed by amination of α-ketoglutarate:

Nitrogen Fixation: Microorganisms Use ATP and a Powerful Reductant to Reduce Atmospheric Nitrogen to Ammonia

Microorganisms use ATP and reduced ferredoxin, a powerful reductant, to reduce N2 to NH3. An iron-molybdenum cluster in nitrogenase deftly catalyzes the fixation of N2, a very inert molecule. Higher organisms consume the fixed nitrogen to synthesize amino acids, nucleotides, and other nitrogen-containing biomolecules. The major points of entry of NH4+ into metabolism are glutamine or glutamate.

Amino Acids are Made from Intermediates of the Citric Acid Cycle and Other Major Pathways

Of the basic set of 20 amino acids (not counting selenocysteine) there are 8 that human beings cannot synthesize. In addition, the amino acids arginine, cysteine, glycine, glutamine, histidine, proline, serine and tyrosine are considered conditionally essential, meaning they are not normally required in the diet, but must be supplied exogenously to specific populations that do not synthesize it in adequate amounts. (For example, enough arginine is synthesized by the urea cycle to meet the needs of an adult but perhaps not those of a growing child.)

Amino acids that need to be obtained from the diet are called essential amino acids. Nonessential amino acids are produced in the body. The pathways for the synthesis of nonessential amino acids are quite simple. Glutamate dehydrogenase catalyzes the reductive amination of α-ketoglutarate to glutamate. A transamination reaction takes place in the synthesis of most amino acids. At this step, the chirality of the amino acid is established. Alanine and aspartate are synthesized by the transamination of pyruvate and oxaloacetate, respectively. Glutamine is synthesized from NH4+ and glutamate, and asparagine is synthesized similarly. Proline and arginine are derived from glutamate. Serine, formed from 3-phosphoglycerate, is the precursor of glycine and cysteine. Tyrosine is synthesized by the hydroxylation of phenylalanine, an essential amino acid. The pathways for the biosynthesis of essential amino acids are much more complex than those for the nonessential ones.

Tetrahydrofolate, a carrier of activated one-carbon units, plays an important role in the metabolism of amino acids and nucleotides. This coenzyme carries one-carbon units at three oxidation states, which are interconvertible: most reduced—methyl; intermediate—methylene; and most oxidized—formyl, formimino, and methenyl. The major donor of activated methyl groups is S-adenosylmethionine, which is synthesized by the transfer of an adenosyl group from ATP to the sulfur atom of methionine. S-Adenosylhomocysteine is formed when the activated methyl group is transferred to an acceptor. It is hydrolyzed to adenosine and homocysteine, the latter of which is then methylated to methionine to complete the activated methyl cycle. The A Cortisol inhibites protein synthesis.

Amino Acid Biosynthesis is Regulated by Feedback Inhibition

Most of the pathways of amino acid biosynthesis are regulated by feedback inhibition, in which the committed step is allosterically inhibited by the final product. Branched pathways require extensive interaction among the branches that includes both negative and positive regulation. The regulation of glutamine synthetase from E. coli is a striking demonstration of cumulative feedback inhibition and of control by a cascade of reversible covalent modifications.

Amino Acids are Precursors of Many Biomolecules

Amino acids are precursors of a variety of biomolecules. Glutathione serves as a sulfhydryl buffer and detoxifying agent. Glutathione peroxidase, a selenoenzyme, catalyzes the reduction of hydrogen peroxide and organic peroxides by glutathione. Nitric oxide, a short-lived messenger, is formed from arginine. Porphyrins are synthesized from glycine and succinyl CoA, which condense to give β-aminolevulinate. Two molecules of this intermediate become linked to form porphobilinogen. Four molecules of porphobilinogen combine to form a linear tetrapyrrole, which cyclizes to uroporphyrinogen III. Oxidation and side-chain modifications lead to the synthesis of protoporphyrin IX, which acquires an iron atom to form heme.

Transcription

In transcription an mRNA chain is generated, with one strand of the DNA double helix in the genome as template. This strand is called the template strand. Transcription can be divided into 3 stages: Initiation, Elongation and Termination, each regulated by a large number of proteins such as transcription factors and coactivators that ensure the correct gene is transcribed.

The DNA strand is read in the 3' to 5' direction and the mRNA is transcribed in the 5' to 3' direction by the RNA polymerase.

Transcription occurs in the cell nucleus, where the DNA is held. The DNA structure of the cell is made up of two helixes made up of sugar and phosphate held together by the bases. The sugar and the phosphate are joined together by covalent bond. The DNA is "unzipped" by the enzyme helicase, leaving the single nucleotide chain open to be copied. RNA polymerase reads the DNA strand from 3 prime (3') end to the 5 prime (5') end, while it synthesizes a single strand of messenger RNA in the 5' to 3' direction. The general RNA structure is very similar to the DNA structure, but in RNA the nucleotide uracil takes the place that thymine occupies in DNA. The single strand of mRNA leaves the nucleus through nuclear pores, and migrates into the cytoplasm.

The first product of transcription differs in prokaryotic cells from that of eukaryotic cells, as in prokaryotic cells the product is mRNA, which needs no post-transcriptional modification, while in eukaryotic cells, the first product is called primary transcript, that needs post-transcriptional modification (capping with 7 methyl guanosine, tailing with a poly A tail) to give hnRNA (heterophil nuclear RNA). hnRNA then undergoes splicing of introns (non coding parts of the gene) via spliceosomes to produce the final mRNA.

Translation

The synthesis of proteins is known as translation. Translation occurs in the cytoplasm, where the ribosomes are located. Ribosomes are made of a small and large subunit that surround the mRNA. In translation, messenger RNA (mRNA) is decoded to produce a specific polypeptide according to the rules specified by the trinucleotide genetic code.

This uses an mRNA sequence as a template to guide the synthesis of a chain of amino acids that form a protein. Translation proceeds in four phases: activation, initiation, elongation, and termination (all describing the growth of the amino acid chain, or polypeptide that is the product of translation).

In activation, the correct amino acid (AA) is joined to the correct transfer RNA (tRNA). While this is not technically a step in translation, it is required for translation to proceed. The AA is joined by its carboxyl group to the 3' OH of the tRNA by an ester bond.

When the tRNA has an amino acid linked to it, it is termed "charged". Initiation involves the small subunit of the ribosome binding to 5' end of mRNA with the help of initiation factors (IF), other proteins that assist the process. Elongation occurs when the next aminoacyl-tRNA (charged tRNA) in line binds to the ribosome along with GTP and an elongation factor.

Termination of the polypeptide happens when the A site of the ribosome faces a stop codon (UAA, UAG, or UGA). When this happens, no tRNA can recognize it, but releasing factor can recognize nonsense codons and causes the release of the polypeptide chain. The capacity of disabling or inhibiting translation in protein biosynthesis is used by antibiotics such as: anisomycin, cycloheximide, chloramphenicol, tetracycline, streptomycin, erythromycin, puromycin etc.

Translation is the process of converting the mRNA codon sequences into an amino acid polypeptide chain.

1. Amino acid activation
2. Initiation - A ribosome attaches to the mRNA and starts to code at the FMet codon (usually AUG, sometimes GUG or UUG).
3. Elongation - tRNA brings the corresponding amino acid (which has an anticodon that identifies the amino acid as the corresponding molecule to a codon) to each codon as the ribosome moves down the mRNA strand.
4. Termination - Reading of the final mRNA codon (aka the STOP codon), which ends the synthesis of the peptide chain and releases it.

Events Following Protein Translation

The events following biosynthesis include post-translational modification and protein folding. During and after synthesis, polypeptide chains often fold to assume, so called, native secondary and tertiary structures. This is known as *protein folding*.

Many proteins undergo *post-translational modification*. This may include the formation of disulfide bridges or attachment of any of a number of biochemical functional groups, such as acetate, phosphate, various lipids and carbohydrates. Enzymes may also remove one or more amino acids from the leading (amino) end of the polypeptide chain, leaving a protein consisting of two polypeptide chains connected by disulfide bonds.

In general, protein molecules are believed to be modified by small chemical groups, post-translationally. Chemical modifications such as, phosphorylation of serine/threonine, acetylation or methylation of lysine, hydroxylation of proline/lysine, formylation of glycine, glycosylation of serine/threonine/ asparagine, acylation of cysteine, myristoylation of glycine, biotinylation of lysine, ubiquitination, etc. on proteins is a very important issue in relation to properly understanding the biological functions of a given protein. These much-studied post-translational modifications have become well-established with the discovery of the respective enzymes (kinases for phosphorylation, acetylases for acetylation, methyl transferases for methylation, etc.), which carry on the chemical modifications on the specific amino acid residues. All of these modifications are still believed to have happened as post-translational events. There is no study yet on when actually one particular modification occurs on a given amino acid residue in a given protein. Does it happen when the protein is already formed, or when the amino acid chain is being synthesized, or before the translation of the primary chain has begun?

Since these chemical modifications are related to the biological functions of a protein, it is easy to think that these chemical modifications have happened to the whole protein molecule, after the protein primary chain is fully synthesized; but, if that is the case, we have to consider the fact that the primary chains get folded instantly, (in a similar way as the newly synthesized DNA strands form helixes), to attain its compact-globular conformation ; As most of the primary chains are fairly long [a 5Kd protein may have 40-45 amino acid residues in its primary chain], it is likely that the newly formed amino acid chain tries to remain intact by folding, thereby avoiding its breakdown via lots of proteases present within the cytoplasm. And, no capping event to protect the N-terminal end of the primary sequence (similar to 5' m-RNA capping to protect m-RNAs) is ever discovered for protein primary structure. So, by folding mechanism, the primary chain, perhaps, avoids the protease attacks.

However, once it gets folded, it may be very difficult for the respective enzyme molecule to find out the particular aa residue from the complexity of that compactly folded conformation. In addition, it can be clearly imagined that this enzymatic modification/reaction on a given amino acid requires presence and association of the appropriate enzyme, necessary cofactors, etc. This association is much easier to occur when the amino acid residues in the primary structure are readily available for binding; in other words, it is much difficult for the enzyme molecules to find and to bind to its substrate amino acid residue in a mature protein molecule after its three-dimensional conformation been attained.

So one can think that the modifications can happen while the primary chain of the protein is being synthesized during the translation process on the m-RNA strand; the amino acid residues on the primary chain can be modified instantly and enzymatically by kinases, acetylases, hydroxylases, methyltransferases, etc. to initiate proper folding for protection in order to avoid degradation by proteases, thereby gaining the globular form.

Also, the reader can imagine another scenario in which, while the free amino acid molecules are formed within a cell and become available, they [in that free state] may be modified enzymatically before taking the ride to the translational event; this means that, while the primary chain is being synthesized, the pre-modified amino acid molecules are ready to be engaged. This way, as the primary chain is getting synthesized, it does not have to be modified by the modifying enzymes anymore, and can fold itself instantaneously without concern of degradation. This also arises new thoughts that (i) phosphate, acetyl, methyl, biotinyl, acyl, etc. groups may have the ability to inhibit protease actions on the primary chains; (ii) most or all of the amino acid residues get modified by small chemical groups (so far, only some chemical groups are known). It is still not fully known exactly when and how the actual modification of a given amino acid residue occurs, at which stage of synthesis, within a protein molecule.

Once the chemically-modified, protease-insensitive, intact protein molecule is generated, it must perform its biological function, which requires its being activated. The activation is probably done by a second set of enzymatic reactions when these chemical groups are removed from the a residues (or added back). So, the second type of post-translational modifications are the opposite reactions of the above-described type, which are dephosphorylation by phosphatases or phosphoryl transferases, deacetylation by deacetylases or acetyltransferases, demethylation by demethylases or methyl-transferases, ubiquitination, SuMoylation, glycosylation, biotinylation, etc. These are taking place on the whole protein molecule toward generating its activated form to execute a particular function or toward its deactivation followed by its total degradation. As the chemical groups are removed (or added), it is quite clear that the proteins can go through different states of structural/conformational change. Thus, after translation, a protein can change its conformation dynamically while the chemical groups are removed or added enzymatically; these conformational changes in the protein structure help the protein to proceed through its lifecycle, until it is ubiquitinated for its total degradation.

6

Organisation of Genome

Genome organization refers to the sequential, not the structural organization of the genome. Besides the coding exons, the non-coding DNA in Eukaryotes may fall in the following classes:

- Introns. They are DNA sequences inserted between the exons and found in the ORF. They are spliced after the first level of transcription. Most introns are junk inserted within genes.
- Pseudogenes. 'Dead', non-functional copies of genes present elsewhere in the genome, but no longer of any use.
- Retropseudogenes. Like pseudogenes, but have been processed, i.e. lack introns. Produced by the action of reverse transcriptase (RT) on mRNA, and subsequent incorporation of the cDNA into the genome.
- Transposons. Jumping genes, which splice themselves in and out of the genome (in DNA form) randomly, by the action of transposase.
- Retrotransposons. Transcribed into an mRNA, which encodes an RT enzyme, which then copies the mRNA back to DNA and incorporates it into the genome.

Infact in humans only 1.5% of the entire genome length corresponds to coding DNA. This 1.5% codes for about 27,000 genes which in turn code for proteins that are responsible for all the cellular processes.

Genome Analysis?

Genome analysis entails the prediction of genes in uncharacterized genomic sequences. The 21st century has seen the announcement of the draft version of the human genome sequence. Model organisms have been sequenced in both the plant and animal kingdoms.

However, the pace of genome annotation is not matching the pace of genome sequencing. Experimental genome annotation is slow and time consuming. The demand is to be able to develop computational tools for gene prediction. Computational Gene prediction is relatively simple for the prokaryotes where all the genes are converted into the corresponding mRNA and then into proteins. The process is more complex for eukaryotic cells where the coding DNA sequence is interrupted by random sequences called introns.

Some of the questions which biologists want to answer today are:

- Given a DNA sequence, what part of it codes for a protein and what part of it is junk DNA.
- Classify the junk DNA as intron, untranslated region, transposons, dead genes, regulatory elements etc.
- Divide a newly sequenced genome into the genes (coding) and the non-coding regions.

The importance of genome analysis can be understood by comparing the human and chimpanzee genomes. The chimp and human genomes vary by an average of just 2% i.e. just about 160 enzymes.

A complete genome analysis of the two genomes would give a strong insight into the various mechanisms responsible for the differences.

Arabidopsis and Humans have the same number of genes, though the Arabidopsis genome is around 250 times smaller than humans. How is that ?

The human genome has a lot of junk DNA, specifically transposons and mobile genetic elements. This increases the size of the human genome, though the number of genes is only 27,000.

However, the number of protein products in humans is significantly higher. Many of the sequenced human genes have alternative splice products. In addition, several other processes (e.g. signal transduction) proceed via further protein modifications, such as Glycosylation. Therefore, the number of human protein products could far exceed the number of genes.

Why do plants have such bulky genomes when they are not as complex as some of the higher eukaryotes ?

This is mostly due to two factors: the ability of plants to duplicate their genomes in order to reproduce (a process known as polyploidization) and the susceptibility of plants to mobile genetic elements.

- Polyploidization allows plants to more easily form hybrids when pollen and ova from different species come together. The result of such hybridization events are plants with genomes that are the sum of the two parent genome sizes (as opposed to half of one parent's genome and half of the other parent's genome as in normal sexual reproduction.
- Also, in case of plants, it is fairly common to observe insertion of transposable elements in intergenic regions. This also explains the difference in the sizes of plant genomes among themselves as well.

What is Open Reading Frame (ORF) ?

The region of the nucleotide sequences from the start codon (ATG) to the stop codon is called the Open Reading frame.

Gene finding in organism specially prokaryotes starts form searching for an open reading frames (ORF). An ORF is a sequence of DNA that starts with start codon "ATG" (not always) and ends with any of the three termination codons (TAA, TAG, TGA). Depending on the starting point, there are six possible ways (three on forward strand and three on complementary strand) of translating any nucleotide sequence into amino acid sequence according to the genetic code.These are called reading frames.

While eukaryotic gene finding is altogether a different task as the eukaryotic genes are not continuous and interrupted by intervening noncoding sequences called 'introns'. Moreover organization of genetic information in eukaryotes and prokaryotes is different.

What is Coding Sequence(CDS)? How is it Different from the ORF?

The Coding Sequence (CDS) is the actual region of DNA that is translated to form proteins. While the ORF may contain introns as well, the CDS refers to those nucleotides(concatenated exons) that can be divided into codons which are actually translated into amino acids by the ribosomal translation machinery. In Prokaryotes the ORF and the CDS are the same.

DNA Replication

Before a cell divides, its DNA is replicated (duplicated.) Because the two strands of a DNA molecule have complementary base pairs, the nucleotide sequence of each strand automatically supplies the information needed to produce its partner.

If the two strands of a DNA molecule are separated, each can be used as a pattern or template to produce a complementary strand. Each template and its new complement together then form a new DNA double helix, identical to the original.

Before replication can occur, the length of the DNA double helix about to be copied must be unwound. In addition, the two strands must be separated, much like the two sides of a zipper, by breaking the weak hydrogen bonds that link the paired bases. Once the DNA strands have been unwound, they must be held apart to expose the bases so that new nucleotide partners can hydrogen-bond to them. The enzyme DNA polymerase then moves along the exposed DNA strand, joining newly arrived nucleotides into a new DNA strand that is complementary to the template. Each cell contains a family of more than thirty enzymes to insure the accurate replication of DNA.

Primers

Though DNA polymerase can elongate a polynucleotide strand by adding new nucleotides, it cannot start a strand from scratch because it can only bond new nucleotides to a free sugar (3') end of a nucleotide chain. DNA polymerase requires the assistance of a primer, a previously existing short strand of DNA (or RNA) that is complementary to the first part of the DNA segment being copied. This small strand of nucleotides anneals (binds) by complementary base pairing to the beginning of the area being copied. With the primer in place, DNA polymerase is then able to continue adding the rest of the pairs of the segment until a new double strand of DNA is completed. Primers are formed from free nucleotides in the cell by enzymes called DNA primases.

Replication Occurs Differently on Antiparallel Strands of DNA

That nucleotides can be added only to the sugar or 3' end of the growing complementary chain presents no problem for the side of the DNA chain opening at its phosphate or 5' end. The primer that binds to the first few exposed bases will end with a sugar (3') where the phosphate of a new nucleotide can be attached. From there on, DNA polymerase can continuously synthesize the growing complementary strand. This strand of DNA is called the leading strand.

A different challenge faces DNA polymerase when the complementary side of the DNA molecule begins unzipping from its sugar (3') toward its phosphate (5') end. A primer of complementary molecules attaching to the opening end of this chain would have a phosphate not a sugar at its exposed end so that new nucleotides could not be joined.

To get around this problem, this strand is synthesized in small pieces backward from the overall direction of replication. This strand is called the lagging strand. The short segments of newly assembled DNA from which the lagging strand is built are called Okazaki fragments. As replication proceeds and nucleotides are added to the 3' end of the Okazaki fragments, they come to meet each other. The primer fragments are then booted out by enzymes and replaced by appropriate DNA nucleotides. The whole thing is then stitched together by another enzyme called DNA ligase.

DNA Damage, Repair, and Recombination

DNA repair refers to a collection of processes by which a cell identifies and corrects damage to the DNA molecules that encode its genome. In human cells, both normal metabolic activities and environmental factors such as UV light and radiation can cause DNA damage, resulting in as many as 1 million individual molecular lesions per cell per day.

Many of these lesions cause structural damage to the DNA molecule and can alter or eliminate the cell's ability to transcribe the gene that the affected DNA encodes. Other lesions induce potentially harmful mutations in the cell's genome, which affect the survival of its daughter cells after it undergoes mitosis. As a consequence, the DNA repair process is constantly active as it responds to damage in the DNA structure. When normal repair processes fail, and when cellular apoptosis does not occur, irreparable DNA damage may occur, including double-strand breaks and DNA crosslinkages. The rate of DNA repair is dependent on many factors, including the cell type, the age of the cell, and the extracellular environment. A cell that has accumulated a large amount of DNA damage, or one that no longer effectively repairs damage incurred to its DNA, can enter one of three possible states:

1. an irreversible state of dormancy, known as senescence
2. cell suicide, also known as apoptosis or programmed cell death
3. unregulated cell division, which can lead to the formation of a tumor that is cancerous.

The DNA repair ability of a cell is vital to the integrity of its genome and thus to its normal functioning and that of the organism. Many genes that were initially shown to influence life span have turned out to be involved in DNA damage repair and protection. Failure to correct molecular lesions in cells that form gametes can introduce mutations into the genomes of the offspring and thus influence the rate of evolution.

DNA Damage

DNA damage, due to environmental factors and normal metabolic processes inside the cell, occurs at a rate of 1,000 to 1,000,000 molecular lesions per cell per day. While this constitutes only 0.000165% of the human genome's approximately 6 billion bases (3 billion base pairs), unrepaired lesions in critical genes (such as tumor suppressor genes) can impede a cell's ability to carry out its function and appreciably increase the likelihood of tumour formation.

The vast majority of DNA damage affects the primary structure of the double helix; that is, the bases themselves are chemically modified. These modifications can in turn disrupt the molecules' regular helical structure by introducing non-native chemical bonds or bulky adducts that do not fit in the standard double helix.

Unlike proteins and RNA, DNA usually lacks tertiary structure and therefore damage or disturbance does not occur at that level. DNA is, however, supercoiled and wound around "packaging" proteins called histones (in eukaryotes), and both superstructures are vulnerable to the effects of DNA damage.

Sources of Damage

DNA damage can be subdivided into two main types:

1. endogenous damage such as attack by reactive oxygen species produced from normal metabolic byproducts (spontaneous mutation), especially the process of oxidative deamination
    - also includes replication errors.
2. exogenous damage caused by external agents such as
    - ultraviolet [UV 200-300nm] radiation from the sun
    - other radiation frequencies, including x-rays and gamma rays
    - hydrolysis or thermal disruption
    - certain plant toxins
    - human-made mutagenic chemicals, especially aromatic compounds that act as DNA intercalating agents
    - cancer chemotherapy and radiotherapy
    - viruses.

The replication of damaged DNA before cell division can lead to the incorporation of wrong bases opposite damaged ones. Daughter cells that inherit these wrong bases carry mutations from which the original DNA sequence is unrecoverable (except in the rare case of a back mutation, for example, through gene conversion).

Types of Damage

There are five main types of damage to DNA due to endogenous cellular processes:

1. *oxidation* of bases [e.g. 8-oxo-7,8-dihydroguanine (8-oxoG)] and generation of DNA strand interruptions from reactive oxygen species,

2. *alkylation* of bases (usually methylation), such as formation of 7-methylguanine, 1-methyladenine, 6-O-Methylguanine
3. *hydrolysis* of bases, such as deamination, depurination, and depyrimidination.
4. "bulky adduct formation" (i.e., benzo[a]pyrene diol epoxide-dG adduct)
5. *mismatch* of bases, due to errors in DNA replication, in which the wrong DNA base is stitched into place in a newly forming DNA strand, or a DNA base is skipped over or mistakenly inserted.

Damage caused by exogenous agents comes in many forms. Some examples are:

1. *UV-B light* causes crosslinking between adjacent cytosine and thymine bases creating *pyrimidine dimers*. This is called direct DNA damage.
2. *UV-A light* creates mostly free radicals. The damage caused by free radicals is called indirect DNA damage.
3. *Ionizing radiation* such as that created by radioactive decay or in *cosmic rays* causes breaks in DNA strands. Low-level ionizing radiation may induce irreparable DNA damage (leading to replicational and transcriptional errors needed for neoplasia or may trigger viral interactions) leading to pre-mature aging and cancer.
4. *Thermal disruption* at elevated temperature increases the rate of depurination (loss of purine bases from the DNA backbone) and single-strand breaks. For example, hydrolytic depurination is seen in the thermophilic bacteria, which grow in hot springs at 40-80 °C. The rate of depurination (300 purine residues per genome per generation) is too high in these species to be repaired by normal repair machinery, hence a possibility of an adaptive response cannot be ruled out.
5. *Industrial chemicals* such as vinyl chloride and hydrogen peroxide, and environmental chemicals such as polycyclic hydrocarbons found in smoke, soot and tar create a huge diversity of DNA adducts- ethenobases, oxidized bases, alkylated phosphotriesters and Crosslinking of DNA just to name a few.

UV damage, alkylation/methylation, X-ray damage and oxidative damage are examples of induced damage. Spontaneous damage can include the loss of a base, deamination, sugar ring puckering and tautomeric shift.

Nuclear versus Mitochondrial DNA Damage

In human cells, and eukaryotic cells in general, DNA is found in two cellular locations - inside the nucleus and inside the mitochondria. Nuclear

DNA (nDNA) exists as chromatin during non-replicative stages of the cell cycle and is condensed into aggregate structures known as chromosomes during cell division. In either state the DNA is highly compacted and wound up around bead-like proteins called histones.

Whenever a cell needs to express the genetic information encoded in its nDNA the required chromosomal region is unravelled, genes located therein are expressed, and then the region is condensed back to its resting conformation. Mitochondrial DNA (mtDNA) is located inside mitochondria organelles, exists in multiple copies, and is also tightly associated with a number of proteins to form a complex known as the nucleoid. Inside mitochondria, reactive oxygen species (ROS), or free radicals, byproducts of the constant production of adenosine triphosphate (ATP) via oxidative phosphorylation, create a highly oxidative environment that is known to damage mtDNA. A critical enzyme in counteracting the toxicity of these species is superoxide dismutase, which is present in both the mitochondria and cytoplasm of eukaryotic cells.

Senescence and Apoptosis

Senescence, an irreversible state in which the cell no longer divides, is a protective response to the shortening of the chromosome ends. The telomeres are long regions of repetitive noncoding DNA that cap chromosomes and undergo partial degradation each time a cell undergoes division. In contrast, quiescence is a reversible state of cellular dormancy that is unrelated to genome damage. Senescence in cells may serve as a functional alternative to apoptosis in cases where the physical presence of a cell for spatial reasons is required by the organism, which serves as a "last resort" mechanism to prevent a cell with damaged DNA from replicating inappropriately in the absence of pro-growth cellular signaling. Unregulated cell division can lead to the formation of a tumour, which is potentially lethal to an organism. Therefore, the induction of senescence and apoptosis is considered to be part of a strategy of protection against cancer.

DNA Damage and Mutation

It is important to distinguish between DNA damage and mutation, the two major types of error in DNA. DNA damages and mutation are fundamentally different. Damages are physical abnormalities in the DNA, such as single- and double-strand breaks, 8-hydroxydeoxyguanosine residues, and polycyclic aromatic hydrocarbon adducts. DNA damages can be recognized by enzymes, and, thus, they can be correctly repaired if redundant information, such as the undamaged sequence in the complementary DNA strand or in a homologous chromosome, is available for copying. If a cell retains DNA damage, transcription of a gene can be prevented, and, thus, translation into a protein will also be blocked. Replication may also be blocked and/or the cell may die.

In contrast to DNA damage, a mutation is a change in the base sequence of the DNA. A mutation cannot be recognized by enzymes once the base change is present in both DNA strands, and, thus, a mutation cannot be repaired. At the cellular level, mutations can cause alterations in protein function and regulation. Mutations are replicated when the cell replicates. In a population of cells, mutant cells will increase or decrease in frequency according to the effects of the mutation on the ability of the cell to survive and reproduce. Although distinctly different from each other, DNA damages and mutations are related because DNA damages often cause errors of DNA synthesis during replication or repair; these errors are a major source of mutation.

Given these properties of DNA damage and mutation, it can be seen that DNA damages are a special problem in non-dividing or slowly dividing cells, where unrepaired damages will tend to accumulate over time. On the other hand, in rapidly dividing cells, unrepaired DNA damages that do not kill the cell by blocking replication will tend to cause replication errors and thus mutation. The great majority of mutations that are not neutral in their effect are deleterious to a cell's survival. Thus, in a population of cells comprising a tissue with replicating cells, mutant cells will tend to be lost. However, infrequent mutations that provide a survival advantage will tend to clonally expand at the expense of neighboring cells in the tissue. This advantage to the cell is disadvantageous to the whole organism, because such mutant cells can give rise to cancer. Thus, DNA damages in frequently dividing cells, because they give rise to mutations, are a prominent cause of cancer. In contrast, DNA damages in infrequently dividing cells are likely a prominent cause of aging.

DNA Repair Mechanisms

Cells cannot function if DNA damage corrupts the integrity and accessibility of essential information in the genome (but cells remain superficially functional when so-called "non-essential" genes are missing or damaged). Depending on the type of damage inflicted on the DNA's double helical structure, a variety of repair strategies have evolved to restore lost information. If possible, cells use the unmodified complementary strand of the DNA or the sister chromatid as a template to recover the original information. Without access to a template, cells use an error-prone recovery mechanism known as translesion synthesis as a last resort. Damage to DNA alters the spatial configuration of the helix, and such alterations can be detected by the cell. Once damage is localized, specific DNA repair molecules bind at or near the site of damage, inducing other molecules to bind and form a complex that enables the actual repair to take place.

Direct Reversal

Cells are known to eliminate three types of damage to their DNA by chemically reversing it. These mechanisms do not require a template, since the types of damage they counteract can occur in only one of the four bases. Such direct reversal mechanisms are specific to the type of damage incurred and do not involve breakage of the phosphodiester backbone.

The formation of pyrimidine dimers upon irradiation with UV light results in an abnormal covalent bond between adjacent pyrimidine bases. The photoreactivation process directly reverses this damage by the action of the enzyme photolyase, whose activation is obligately dependent on energy absorbed from blue/UV light (300–500 nm wavelength) to promote catalysis. Another type of damage, methylation of guanine bases, is directly reversed by the protein methyl guanine methyl transferase (MGMT), the bacterial equivalent of which is called ogt. This is an expensive process because each MGMT molecule can be used only once; that is, the reaction is stoichiometric rather than catalytic. A generalized response to methylating agents in bacteria is known as the adaptive response and confers a level of resistance to alkylating agents upon sustained exposure by upregulation of alkylation repair enzymes. The third type of DNA damage reversed by cells is certain methylation of the bases cytosine and adenine.

Single-strand Damage

When only one of the two strands of a double helix has a defect, the other strand can be used as a template to guide the correction of the damaged strand. In order to repair damage to one of the two paired molecules of DNA, there exist a number of excision repair mechanisms that remove the damaged nucleotide and replace it with an undamaged nucleotide complementary to that found in the undamaged DNA strand.

1. Base excision repair (BER), which repairs damage to a single base caused by oxidation, alkylation, hydrolysis, or deamination. The damaged base is removed by a DNA glycosylase. The "missing tooth" is then recognised by an enzyme called AP endonuclease, which cuts the Phosphodiester bond. The missing part is then resynthesized by a DNA polymerase, and a DNA ligase performs the final nick-sealing step.
2. Nucleotide excision repair (NER), which recognizes bulky, helix-distorting lesions such as pyrimidine dimers and 6,4 photoproducts. A specialized form of NER known as transcription-coupled repair deploys NER enzymes to genes that are being actively transcribed.
3. Mismatch repair (MMR), which corrects errors of DNA replication and recombination that result in mispaired (but undamaged) nucleotides.

Double-strand Breaks

Double-strand breaks, in which both strands in the double helix are severed, are particularly hazardous to the cell because they can lead to genome rearrangements. Three mechanisms exist to repair double-strand breaks (DSBs): non-homologous end joining (NHEJ), microhomology-mediated end joining (MMEJ), and homologous recombination. PVN Acharya noted that double-strand breaks and a "cross-linkage joining both strands at the same point is irreparable because neither strand can then serve as a template for repair. The cell will die in the next mitosis or in some rare instances, mutate."

In NHEJ, DNA Ligase IV, a specialized DNA ligase that forms a complex with the cofactor XRCC4, directly joins the two ends. To guide accurate repair, NHEJ relies on short homologous sequences called microhomologies present on the single-stranded tails of the DNA ends to be joined. If these overhangs are compatible, repair is usually accurate. NHEJ can also introduce mutations during repair. Loss of damaged nucleotides at the break site can lead to deletions, and joining of nonmatching termini forms translocations. NHEJ is especially important before the cell has replicated its DNA, since there is no template available for repair by homologous recombination. There are "backup" NHEJ pathways in higher eukaryotes. Besides its role as a genome caretaker, NHEJ is required for joining hairpin-capped double-strand breaks induced during V(D)J recombination, the process that generates diversity in B-cell and T-cell receptors in the vertebrate immune system.

Homologous recombination requires the presence of an identical or nearly identical sequence to be used as a template for repair of the break. The enzymatic machinery responsible for this repair process is nearly identical to the machinery responsible for chromosomal crossover during meiosis. This pathway allows a damaged chromosome to be repaired using a sister chromatid (available in G2 after DNA replication) or a homologous chromosome as a template. DSBs caused by the replication machinery attempting to synthesize across a single-strand break or unrepaired lesion cause collapse of the replication fork and are typically repaired by recombination.

Topoisomerases introduce both single- and double-strand breaks in the course of changing the DNA's state of supercoiling, which is especially common in regions near an open replication fork. Such breaks are not considered DNA damage because they are a natural intermediate in the topoisomerase biochemical mechanism and are immediately repaired by the enzymes that created them.

A team of French researchers bombarded *Deinococcus radiodurans* to study the mechanism of double-strand break DNA repair in that organism. At least two copies of the genome, with random DNA breaks, can form DNA fragments through annealing. Partially overlapping fragments are then used for synthesis of homologous regions through a moving D-loop that can continue extension until they find complementary partner strands. In the final step there is crossover by means of RecA-dependent homologous recombination.

Translesion Synthesis

Translesion synthesis is a DNA damage tolerance process that allows the DNA replication machinery to replicate past DNA lesions such as thymine dimers or AP sites. It involves switching out regular DNA polymerases for specialized translesion polymerases (i.e. DNA polymerase IV or V, from the Y Polymerase family), often with larger active sites that can facilitate the insertion of bases opposite damaged nucleotides. The polymerase switching is thought to be mediated by, among other factors, the post-translational modification of the replication processivity factor PCNA. Translesion synthesis polymerases often have low fidelity (high propensity to insert wrong bases) on undamaged templates relative to regular polymerases. However, many are extremely efficient at inserting correct bases opposite specific types of damage. For example, Pol ç mediates error-free bypass of lesions induced by UV irradiation, whereas Pol æ introduces mutations at these sites. From a cellular perspective, risking the introduction of point mutations during translesion synthesis may be preferable to resorting to more drastic mechanisms of DNA repair, which may cause gross chromosomal aberrations or cell death. In short, the process involves specialized polymerases either bypassing or repairing lesions at locations of stalled DNA replication. A bypass platform is provided to these polymerases by Proliferating cell nuclear antigen (PCNA). Under normal circumstances, PCNA bound to polymerases replicates the DNA. At a site of lesion, PCNA is ubiquitinated, or modified, by the RAD6/RAD18 proteins to provide a platform for the specialized polymerases to bypass the lesion and resume DNA replication.

Global Response to DNA Damage

Cells exposed to ionizing radiation, ultraviolet light or chemicals are prone to acquire multiple sites of bulky DNA lesions and double-strand breaks. Moreover, DNA damaging agents can damage other biomolecules such as proteins, carbohydrates, lipids, and RNA. The accumulation of damage, to be specific, double-strand breaks or adducts stalling the replication forks, are among known stimulation signals for a global response to DNA damage. The global response to damage is an act directed toward the cells' own preservation and triggers multiple pathways of macromolecular repair, lesion bypass, tolerance, or apoptosis. The common features of global response are induction of multiple genes, cell cycle arrest, and inhibition of cell division.

DNA Damage Checkpoints

After DNA damage, cell cycle checkpoints are activated. Checkpoint activation pauses the cell cycle and gives the cell time to repair the damage before continuing to divide. DNA damage checkpoints occur at the G1/S and G2/M boundaries. An intra-S checkpoint also exists. Checkpoint activation is controlled by two master kinases, ATM and ATR. ATM responds to DNA double-strand breaks and disruptions in chromatin structure, whereas ATR primarily responds to stalled replication forks. These kinases phosphorylate downstream targets in a signal transduction cascade, eventually leading to cell cycle arrest. A class of checkpoint mediator proteins including BRCA1, MDC1, and 53BP1 has also been identified. These proteins seem to be required for transmitting the checkpoint activation signal to downstream proteins.

P53 is an important downstream target of ATM and ATR, as it is required for inducing apoptosis following DNA damage. At the G1/S checkpoint, p53 functions by deactivating the CDK2/cyclin E complex. Similarly, p21 mediates the G2/M checkpoint by deactivating the CDK1/cyclin B complex.

The Prokaryotic SOS Response

The SOS response is the term used to describe changes in gene expression in *Escherichia coli* and other bacteria in response to extensive DNA damage. The prokaryotic SOS system is regulated by two key proteins: LexA and RecA. The LexA homodimer is a transcriptional repressor that binds to operator sequences commonly referred to as SOS boxes.

In *Escherichia coli* it is known that LexA regulates transcription of approximately 48 genes including the lexA and recA genes. The SOS response is known to be widespread in the Bacteria domain, but it is mostly absent in some bacterial phyla, like the Spirochetes.

The most common cellular signals activating the SOS response are regions of single-stranded DNA (ssDNA), arising from stalled replication forks or double-strand breaks, which are processed by DNA helicase to separate the two DNA strands. In the initiation step, RecA protein binds to ssDNA in an ATP hydrolysis driven reaction creating RecA–ssDNA filaments. RecA–ssDNA filaments activate LexA autoprotease activity, which ultimately leads to cleavage of LexA dimer and subsequent LexA degradation. The loss of LexA repressor induces transcription of the SOS genes and allows for further signal induction, inhibition of cell division and an increase in levels of proteins responsible for damage processing.

In *Escherichia coli*, SOS boxes are 20-nucleotide long sequences near promoters with palindromic structure and a high degree of sequence conservation. In other classes and phyla, the sequence of SOS boxes varies considerably, with different length and composition, but it is always highly conserved and one of the strongest short signals in the genome. The high information content of SOS boxes permits differential binding of LexA to different promoters and allows for timing of the SOS response.

The lesion repair genes are induced at the beginning of SOS response. The error-prone translesion polymerases, for example, UmuCD'2 (also called DNA polymerase V), are induced later on as a last resort. Once the DNA damage is repaired or bypassed using polymerases or through recombination, the amount of single-stranded DNA in cells is decreased, lowering the amounts of RecA filaments decreases cleavage activity of LexA homodimer, which then binds to the SOS boxes near promoters and restores normal gene expression.

Eukaryotic Transcriptional Responses to DNA Damage

Eukaryotic cells exposed to DNA damaging agents also activate important defensive pathways by inducing multiple proteins involved in DNA repair, cell cycle checkpoint control, protein trafficking and degradation. Such genome wide transcriptional response is very complex and tightly regulated, thus allowing coordinated global response to damage.

Exposure of yeast *Saccharomyces cerevisiae* to DNA damaging agents results in overlapping but distinct transcriptional profiles. Similarities to environmental shock response indicates that a general global stress response pathway exist at the level of transcriptional activation. In contrast, different human cell types respond to damage differently indicating an absence of a common global response.

The probable explanation for this difference between yeast and human cells may be in the heterogeneity of mammalian cells. In an animal different types of cells are distributed among different organs that have evolved different sensitivities to DNA damage.

In general global response to DNA damage involves expression of multiple genes responsible for postreplication repair, homologous recombination, nucleotide excision repair, DNA damage checkpoint, global transcriptional activation, genes controlling mRNA decay, and many others. A large amount of damage to a cell leaves it with an important decision: undergo apoptosis and die, or survive at the cost of living with a modified genome.

An increase in tolerance to damage can lead to an increased rate of survival that will allow a greater accumulation of mutations. Yeast Rev1 and human polymerase ç are members of [Y family translesion DNA polymerases present during global response to DNA damage and are responsible for enhanced mutagenesis during a global response to DNA damage in eukaryotes.

DNA Repair and Aging

Pathological Effects of Poor DNA Repair

DNA repair rate is an important determinant of cell pathology.

Experimental animals with genetic deficiencies in DNA repair often show decreased life span and increased cancer incidence. For example, mice deficient in the dominant NHEJ pathway and in telomere maintenance mechanisms get lymphoma and infections more often, and, as a consequence, have shorter lifespans than wild-type mice. In similar manner, mice deficient in a key repair and transcription protein that unwinds DNA helices have premature onset of aging-related diseases and consequent shortening of lifespan. However, not every DNA repair deficiency creates exactly the predicted effects; mice deficient in the NER pathway exhibited shortened life span without correspondingly higher rates of mutation.

If the rate of DNA damage exceeds the capacity of the cell to repair it, the accumulation of errors can overwhelm the cell and result in early senescence, apoptosis, or cancer. Inherited diseases associated with faulty DNA repair functioning result in premature aging, increased sensitivity to carcinogens, and correspondingly increased cancer risk. On the other hand, organisms with enhanced DNA repair systems, such as *Deinococcus radiodurans*, the most radiation-resistant known organism, exhibit remarkable resistance to the double-strand break-inducing effects of radioactivity, likely due to enhanced efficiency of DNA repair and especially NHEJ.

Longevity and Caloric Restriction

A number of individual genes have been identified as influencing variations in life span within a population of organisms. The effects of these genes is strongly dependent on the environment, in particular, on the organism's diet. Caloric restriction reproducibly results in extended lifespan in a variety of organisms, likely via nutrient sensing pathways and decreased metabolic rate. The molecular mechanisms by which such restriction results in lengthened lifespan are as yet unclear; however, the behavior of many genes known to be involved in DNA repair is altered under conditions of caloric restriction.

For example, increasing the gene dosage of the gene SIR-2, which regulates DNA packaging in the nematode worm *Caenorhabditis elegans*, can significantly extend lifespan. The mammalian homolog of SIR-2 is known to induce downstream DNA repair factors involved in NHEJ, an activity that is especially promoted under conditions of caloric restriction. Caloric restriction has been closely linked to the rate of base excision repair in the nuclear DNA of rodents, although similar effects have not been observed in mitochondrial DNA.

It is interesting to note that the *C. elegans* gene AGE-1, an upstream effector of DNA repair pathways, confers dramatically extended life span under free-feeding conditions but leads to a decrease in reproductive fitness under conditions of caloric restriction. This observation supports the pleiotropy theory of the biological origins of aging, which suggests that genes conferring a large survival advantage early in life will be selected for even if they carry a corresponding disadvantage late in life.

Medicine and DNA Repair Modulation

Hereditary DNA Repair Disorders: Defects in the NER mechanism are responsible for several genetic disorders, including:

- Xeroderma pigmentosum: hypersensitivity to sunlight/UV, resulting in increased skin cancer incidence and premature aging
- Cockayne syndrome: hypersensitivity to UV and chemical agents
- Trichothiodystrophy: sensitive skin, brittle hair and nails.

Mental retardation often accompanies the latter two disorders, suggesting increased vulnerability of developmental neurons.

Other DNA repair disorders include:

- Werner's syndrome: premature aging and retarded growth
- Bloom's syndrome: sunlight hypersensitivity, high incidence of malignancies (especially leukemias).
- Ataxia telangiectasia: sensitivity to ionizing radiation and some chemical agents.

All of the above diseases are often called "segmental progerias" ("accelerated aging diseases") because their victims appear elderly and suffer from aging-related diseases at an abnormally young age, while not manifesting all the symptoms of old age.

Other diseases associated with reduced DNA repair function include Fanconi's anemia, hereditary breast cancer and hereditary colon cancer.

DNA Repair and Cancer

Inherited mutations that affect DNA repair genes are strongly associated with high cancer risks in humans. Hereditary nonpolyposis colorectal cancer (HNPCC) is strongly associated with specific mutations in the DNA mismatch repair pathway. BRCA1 and BRCA2, two famous mutations conferring a hugely increased risk of breast cancer on carriers, are both associated with a large number of DNA repair pathways, especially NHEJ and homologous recombination.

Cancer therapy procedures such as chemotherapy and radiotherapy work by overwhelming the capacity of the cell to repair DNA damage, resulting in cell death. Cells that are most rapidly dividing - most typically cancer cells - are preferentially affected. The side-effect is that other non-cancerous but rapidly dividing cells such as stem cells in the bone marrow are also affected. Modern cancer treatments attempt to localize the DNA damage to cells and tissues only associated with cancer, either by physical means (concentrating the therapeutic agent in the region of the tumour) or by biochemical means (exploiting a feature unique to cancer cells in the body).

DNA Repair and Evolution

The basic processes of DNA repair are highly conserved among both prokaryotes and eukaryotes and even among bacteriophage (viruses that infect bacteria); however, more complex organisms with more complex genomes have correspondingly more complex repair mechanisms. The ability of a large number of protein structural motifs to catalyze relevant chemical reactions has played a significant role in the elaboration of repair mechanisms during evolution. For an extremely detailed review of hypotheses relating to the evolution of DNA repair.

The fossil record indicates that single-cell life began to proliferate on the planet at some point during the Precambrian period, although exactly when recognizably modern life first emerged is unclear. Nucleic acids became the sole and universal means of encoding genetic information, requiring DNA repair mechanisms that in their basic form have been inherited by all extant life forms from their common ancestor. The emergence of Earth's oxygen-rich atmosphere (known as the "oxygen catastrophe") due to photosynthetic organisms, as well as the presence of potentially damaging free radicals in the cell due to oxidative phosphorylation, necessitated the evolution of DNA repair mechanisms that act specifically to counter the types of damage induced by oxidative stress.

Rate of Evolutionary Change

On some occasions, DNA damage is not repaired, or is repaired by an error-prone mechanism that results in a change from the original sequence. When this occurs, mutations may propagate into the genomes of the cell's progeny. Should such an event occur in a germ line cell that will eventually produce a gamete, the mutation has the potential to be passed on to the organism's offspring. The rate of evolution in a particular species (or, in a particular gene) is a function of the rate of mutation. As a consequence, the rate and accuracy of DNA repair mechanisms have an influence over the process of evolutionary change.

Senescence

Senescence or biological aging is the change in the biology of an organism as it ages after its maturity. Such changes range from those affecting its cells and their function to that of the whole organism. There are a number of theories as to why senescence occurs, including ones that claim it is programmed by gene expression changes and that it is the accumulative damage of biological processes.

The word *senescence* is derived from the Latin word *senex*, meaning *old man*, *old age*, or *advanced in age*.

Cellular Senescence

Cellular senescence is the phenomenon by which normal diploid cells lose the ability to divide, normally after about 50 cell divisions in vitro. Some cells become senescent after fewer replications cycles as a result of DNA double strand breaks, toxins, etc. This phenomenon is also known as "replicative senescence", the "Hayflick phenomenon", or the Hayflick limit in honour of Dr. Leonard Hayflick who was the first to publish this information in 1965. In response to DNA damage (including shortened telomeres), cells either age or self-destruct (apoptosis, programmed cell death) if the damage cannot be repaired. In this 'cellular suicide', the death of one cell, or more, may benefit the organism as a whole. For example, in plants the death of the water-conducting xylem cells (tracheids and vessel elements) allows the cells to function more efficiently and so deliver water to the upper parts of a plant.

Aging of the Whole Organism

Organismal senescence is the aging of whole organisms. In general, aging is characterized by the declining ability to respond to stress, increased homeostatic imbalance, and increased risk of aging-associated diseases. Death is the ultimate consequence of aging, though "old age" is not a scientifically recognized cause of death because there is always a specific proximal cause, such as cancer, heart disease, or liver failure. Aging of whole organisms is therefore a complex process that can be defined as "a progressive deterioration of physiological function, an intrinsic age-related process of loss of viability and increase in vulnerability". Differences in maximum life span among species correspond to different "rates of aging". For example, inherited differences in the rate of aging make a mouse elderly at 3 years and a human elderly at 80 years. These genetic differences affect a variety of physiological processes, including the efficiency of DNA repair, antioxidant enzymes, and rates of free radical production. Senescence of the organism gives rise to the Gompertz–Makeham law of mortality, which says that mortality rate rises rapidly with age.

Some animals, such as some reptiles and fish, age slowly (negligible senescence) and exhibit very long lifespans. Some even exhibit "negative senescence", in which mortality falls with age, in disagreement with the Gompertz–Makeham "law". Whether replicative senescence (Hayflick limit) plays a causative role in organismal aging is at present an active area of investigation.

Theories of Aging

The process of senescence is complex, and may derive from a variety of different mechanisms and exist for a variety of different reasons. However, senescence is not universal, and scientific evidence suggests that cellular senescence evolved in certain species because it prevents the onset of cancer. In a few simple species, such as Hydra, senescence is negligible and cannot be detected. All such species have no "post-mitotic" cells; they reduce the effect of damaging free radicals by cell division and dilution. Such species are not immortal, however, as they will eventually fall prey to trauma or disease. Moreover, average lifespans can vary greatly within and between species. This suggests that both genetic and environmental factors contribute to aging.

In general, theories that explain senescence have been divided between the programmed and stochastic theories of aging. Programmed theories imply that aging is regulated by biological clocks operating throughout the lifespan. This regulation would depend on changes in gene expression that affect the systems responsible for maintenance, repair, and defense responses. Stochastic theories blame environmental impacts on living organisms that induce cumulative damage at various levels as the cause of aging, examples of which ranging from damage to DNA, damage to tissues and cells by oxygen radicals (widely known as free radicals countered by the even more well-known antioxidants), and cross-linking.

However, aging is seen as a progressive failure of homeodynamics (homeostasis) involving genes for the maintenance and repair, stochastic events leading to molecular damage and molecular heterogeneity, and chance events determining the probability of death. Since complex and interacting systems of maintenance and repair comprise the homeodynamic (old term: homeostasis) space of a biological system, aging is considered to be a progressive shrinkage of homeodynamic space mainly due to increased molecular heterogeneity.

Evolutionary Theories

A gene can be expressed at various life-stages. Therefore, natural selection can support lethal and harmful alleles, if their expression occurs after reproduction. Senescence may be the product of such selection. In addition, aging is believed to have evolved because of the increasingly smaller probability of an organism still being alive at older age, due to predation and accidents, both of which may be random and age-invariant. It is thought that strategies that result in a higher reproductive rate at a young age, but shorter overall lifespan, result in a higher lifetime reproductive success and are therefore favoured by natural selection. In essence, aging is, therefore, the result of investing resources in reproduction, rather than maintenance of the body (the "Disposable Soma" theory), in light of the fact that accidents, predation, and disease will eventually kill the organism no matter how much energy is devoted to repair of the body. Various other theories of aging exist, and are not necessarily mutually exclusive.

The geneticist J. B. S. Haldane wondered why the dominant mutation that causes Huntington's disease remained in the population, and why natural selection had not eliminated it. The onset of this neurological disease is (on average) at age 45 and is invariably fatal within 10–20 years. Haldane assumed that, in human prehistory, few survived until age 45. Since few were alive at older ages and their contribution to the next generation was therefore small relative to the large cohorts of younger age groups, the force of selection against such late-acting deleterious mutations was correspondingly small. However, if a mutation affected younger individuals, selection against it would be strong. Therefore, late-acting deleterious mutations could accumulate in populations over evolutionary time through genetic drift. This principle has been demonstrated experimentally. And it is these later-acting deleterious mutations that are believed to cause—even allow—age-related mortality.

Peter Medawar formalised this observation in his mutation accumulation theory of aging. "The force of natural selection weakens with increasing age—even in a theoretically immortal population, provided only that it is exposed to real hazards of mortality. If a genetic disaster... happens late enough in individual life, its consequences may be completely unimportant". The 'real hazards of mortality' are, in typical circumstances, predation, disease, and accidents. So, even an immortal population, whose fertility does not decline with time, will have fewer individuals alive in older age groups. This is called 'extrinsic mortality'. Young cohorts, not depleted in numbers yet by extrinsic mortality, contribute far more to the next generation than the few remaining older cohorts, so the force of selection against late-acting deleterious mutations, which affect only these few older individuals, is very weak. The mutations may not be selected against, therefore, and may spread over evolutionary time into the population.

The major testable prediction made by this model is that species that have high extrinsic mortality in nature will age more quickly and have shorter intrinsic lifespans. This is borne out among mammals, the best-studied in terms of life history. There is a correlation among mammals between body size and lifespan, such that larger species live longer than smaller species in controlled/optimum conditions, but there are notable exceptions. For instance, many bats and rodents are of similar size, yet bats live much longer. For instance, the little brown bat, half the size of a mouse, can live 30 years in the wild. A mouse will live 2–3 years even with optimum conditions. The explanation is that bats have fewer predators, so therefore low extrinsic mortality. Thus, more individuals survive to later ages, so the force of selection against late-acting deleterious mutations is stronger. Fewer late-acting deleterious mutations = slower aging = longer lifespan. Birds are also warm-blooded and are of similar size to many small mammals, yet live often 5–10 times as long. They have fewer predation pressures compared with ground-dwelling mammals. And seabirds, which, in general, have the fewest predators of all birds, live longest.

Also, when examining the body-size vs. lifespan relationship, one will observe that predator mammals tend to have longer lifespans than prey animals in a controlled environment such as a zoo or nature reserve. The explanation for the long lifespans of primates (such as humans, monkeys, and apes) relative to body size is that their intelligence and often sociality helps them avoid becoming prey. Being a predator, being smart, and working together all reduce extrinsic mortality.

Another evolutionary theory of aging was proposed by George C. Williams and involves antagonistic pleiotropy. A single gene may affect multiple traits. Some traits that increase fitness early in life may also have negative effects later in life. But, because many more individuals are alive at young ages than at old ages, even small positive effects early can be strongly selected for, and large negative effects later may be very weakly selected against. Williams suggested the following example: Perhaps a gene codes for calcium deposition in bones, which promotes juvenile survival and will therefore be favored by natural selection; however, this same gene promotes calcium deposition in the arteries, causing negative effects in old age. Therefore, negative effects in old age may reflect the result of natural selection for pleiotropic genes that are beneficial early in life. In this case, fitness is relatively high when Fisher's reproductive value is high and relatively low when Fisher's reproductive value is low.

Gene Regulation

A number of genetic components of aging have been identified using model organisms, ranging from the simple budding yeast *Saccharomyces cerevisiae* to worms such as *Caenorhabditis elegans* and fruit flies (*Drosophila melanogaster*). Study of these organisms has revealed the presence of at least two conserved aging pathways.

One of these pathways involves the gene *Sir2*, a NAD+-dependent histone deacetylase. In yeast, Sir2 is required for genomic silencing at three loci: The yeast mating loci, the telomeres and the ribosomal DNA (rDNA). In some species of yeast, replicative aging may be partially caused by homologous recombination between rDNA repeats; excision of rDNA repeats results in the formation of extrachromosomal rDNA circles (ERCs). These ERCs replicate and preferentially segregate to the mother cell during cell division, and are believed to result in cellular senescence by titrating away (competing for) essential nuclear factors. ERCs have not been observed in other species (nor even all strains of the same yeast species) of yeast (which also display replicative senescence), and ERCs are not believed to contribute to aging in higher organisms such as humans (they have not been shown to accumulate in mammals in a similar manner to yeast). Extrachromosomal circular DNA (eccDNA) has been found in worms, flies, and humans. The origin and role of eccDNA in aging, if any, is unknown.

Despite the lack of a connection between circular DNA and aging in higher organisms, extra copies of Sir2 are capable of extending the lifespan of both worms and flies (though, in flies, this finding has not been replicated by other investigators, and the activator of Sir2 resveratrol does not reproducibly increase lifespan in either species). Whether the Sir2 homologues in higher organisms have any role in lifespan is unclear, but the human SIRT1 protein has been demonstrated to deacetylate p53, Ku70, and the forkhead family of transcription factors. SIRT1 can also regulate acetylates such as CBP/p300, and has been shown to deacetylate specific histone residues.

RAS1 and RAS2 also affect aging in yeast and have a human homologue. RAS2 overexpression has been shown to extend lifespan in yeast. Other genes regulate aging in yeast by increasing the resistance to oxidative stress. Superoxide dismutase, a protein that protects against the effects of mitochondrial free radicals, can extend yeast lifespan in stationary phase when overexpressed. In higher organisms, aging is likely to be regulated in part through the insulin/IGF-1 pathway. Mutations that affect insulin-like signaling in worms, flies, and the growth hormone/IGF1 axis in mice are associated with extended lifespan. In yeast, Sir2 activity is regulated by the nicotinamidase PNC1. PNC1 is transcriptionally upregulated under stressful conditions such as caloric restriction, heat shock, and osmotic shock. By converting nicotinamide to niacin, nicotinamide is removed, inhibiting the activity of Sir2. A nicotinamidase found in humans, known as PBEF, may serve a similar function, and a secreted form of PBEF known as visfatin may help to regulate serum insulin levels. It is not known, however, whether these mechanisms also exist in humans, since there are obvious differences in biology between humans and model organisms.

Sir2 activity has been shown to increase under calorie restriction. Due to the lack of available glucose in the cells, more NAD+ is available and can activate Sir2. Resveratrol, a stilbenoid found in the skin of red grapes, was reported to extend the lifespan of yeast, worms, and flies (the lifespan extension in flies and worms have proved irreproducible by independent investigators). It has been shown to activate Sir2 and therefore mimics the effects of calorie restriction, if one accepts that caloric restriction is indeed dependent on Sir2.

Gene expression is imperfectly controlled, and it is possible that random fluctuations in the expression levels of many genes contribute to the aging process as suggested by a study of such genes in yeast. Individual cells, which are genetically identical, none-the-less can have substantially different responses to outside stimuli, and markedly different lifespans, indicating the epigenetic factors play an important role in gene expression and aging as well as genetic factors.

According to the GenAge database of aging-related genes there are over 700 genes associated with aging in model organisms: 555 in the soil roundworm (*Caenorhabditis elegans*), 87 in the bakers' yeast (*Saccharomyces cerevisiae*), 75 in the fruit fly (*Drosophila melanogaster*) and 68 in the mouse (*Mus musculus*).

Cellular Senescence

As noted above, senescence is not universal, and senescence is not observed in single-celled organisms that reproduce through the process of cellular mitosis. Moreover, cellular senescence is not observed in several organisms, including perennial plants, sponges, corals, and lobsters. In those species where cellular senescence is observed, cells eventually become post-mitotic when they can no longer replicate themselves through the process of cellular mitosis; i.e., cells experience *replicative senescence*. How and why some cells become post-mitotic in some species has been the subject of much research and speculation, but (as noted above) it is widely believed that cellular senescence evolved as a way to prevent the onset and spread of cancer. Somatic cells that have divided many times will have accumulated DNA mutations and would therefore be in danger of becoming cancerous if cell division continued.

Lately, the role of telomeres in cellular senescence has aroused general interest, especially with a view to the possible genetically adverse effects of cloning. The successive shortening of the chromosomal telomeres with each cell cycle is also believed to limit the number of divisions of the cell, thus contributing to aging. There have, on the other hand, also been reports that cloning could alter the shortening of telomeres. Some cells do not age and are, therefore, described as being "biologically immortal". It is theorized by some that when it is discovered exactly what allows these cells, whether it be the result of telomere lengthening or not, to divide without limit that it will be possible to genetically alter other cells to have the same capability. It is further theorized that it will eventually be possible to genetically engineer all cells in the human body to have this capability by employing gene therapy and, therefore, stop or reverse aging, effectively making the entire organism potentially immortal.

Cancer cells are usually immortal. In about 85% of tumors, this evasion of cellular senescence is the result of up-activation of their telomerase genes. This simple observation suggests that reactivation of telomerase in healthy individuals could greatly increase their cancer risk. Whether cell senescence plays any role in organismal aging is at present unknown, and is an active area of investigation. Mouse mutants lacking telomerase do not immediately show accelerated aging.

Chemical Damage

One of the earliest aging theories was the *Rate of Living Hypothesis* described by Raymond Pearl in 1928(based on earlier work by Max Rubner), which states that fast basal metabolic rate corresponds to short maximum life span.

While there may be some validity to the idea that for various types of specific damage detailed below that are by-products of metabolism, all other things being equal, a fast metabolism may reduce lifespan, in general this theory does not adequately explain the differences in lifespan either within, or between, species.

Calorically-restricted animals process as much, or more, calories per gram of body mass, as their *ad libitum* fed counterparts, yet exhibit substantially longer lifespans.. Similarly, metabolic rate is a poor predictor of lifespan for birds, bats and other species that, it is presumed, have reduced mortality from predation, and therefore have evolved long lifespans even in the presence of very high metabolic rates. More recently, it was shown that, when modern statistical methods for correcting for the effects of body size and phylogeny are employed, metabolic rate does not correlate with longevity in mammals or birds.

With respect to specific types of chemical damage caused by metabolism, it is suggested that damage to long-lived biopolymers, such as structural proteins or DNA, caused by ubiquitous chemical agents in the body such as oxygen and sugars, are in part responsible for aging. The damage can include breakage of biopolymer chains, cross-linking of biopolymers, or chemical attachment of unnatural substituents (haptens) to biopolymers.

Under normal aerobic conditions, approximately 4% of the oxygen metabolized by mitochondria is converted to superoxide ion, which can subsequently be converted to hydrogen peroxide, hydroxyl radical and eventually other reactive species including other peroxides and singlet oxygen, which can, in turn, generate free radicals capable of damaging structural proteins and DNA. Certain metal ions found in the body, such as copper and iron, may participate in the process. (In Wilson's disease, a hereditary defect that causes the body to retain copper, some of the symptoms resemble accelerated senescence.) These processes are termed *oxidative damage* and are linked to the benefits of nutritionally derived polyphenol antioxidants.

Sugars such as glucose and fructose can react with certain amino acids such as lysine and arginine and certain DNA bases such as guanine to produce sugar adducts, in a process called *glycation*. These adducts can further rearrange to form reactive species, which can then cross-link the structural proteins or DNA to similar biopolymers or other biomolecules such as non-structural proteins. People with diabetes, who have elevated blood sugar, develop senescence-associated disorders much earlier than the general population, but can delay such disorders by rigorous control of their blood sugar levels. There is evidence that sugar damage is linked to oxidant damage in a process termed *glycoxidation*.

Free radicals can damage proteins, lipids or DNA. Glycation mainly damages proteins. Damaged proteins and lipids accumulate in lysosomes as lipofuscin. Chemical damage to structural proteins can lead to loss of function; for example, damage to collagen of blood vessel walls can lead to vessel-wall stiffness and, thus, hypertension, and vessel wall thickening and reactive tissue formation (atherosclerosis); similar processes in the kidney can lead to renal failure. Damage to enzymes reduces cellular functionality. Lipid peroxidation of the inner mitochondrial membrane reduces the electric potential and the ability to generate energy. It is probably no accident that nearly all of the so-called "accelerated aging diseases" are due to defective DNA repair enzymes. It is believed that the impact of alcohol on aging can be partly explained by alcohol's activation of the HPA axis, which stimulates glucocorticoid secretion, long-term exposure to which produces symptoms of aging.

Reliability Theory

Reliability theory suggests that biological systems start their adult life with a high load of initial damage. Reliability theory is a general theory about systems failure. It allows researchers to predict the age-related failure kinetics for a system of given architecture (reliability structure) and given reliability of its components. Reliability theory predicts that even those systems that composed entirely of non-aging elements (with a constant failure rate) will nevertheless deteriorate (fail more often) with age, if these systems are redundant in irreplaceable elements. Aging, therefore, is a direct consequence of systems. Reliability theory also predicts the late-life mortality deceleration with subsequent leveling-off, as well as the late-life mortality plateaus, as an inevitable consequence of redundancy exhaustion at extreme old ages.

The theory explains why mortality rates increase exponentially with age (the Gompertz law) in many species, by taking into account the initial flaws (defects) in newly formed systems. It also explains why organisms "prefer" to die according to the Gompertz law, while technical devices usually fail according to the Weibull (power) law. Reliability theory allows to specify conditions when organisms die according to the Weibull distribution: Organisms should be relatively free of initial flaws and defects.

The theory makes it possible to find a general failure law applicable to all adult and extreme old ages, where the Gompertz and the Weibull laws are just special cases of this more general failure law. The theory explains why relative differences in mortality rates of compared populations (within a given species) vanish with age (compensation law of mortality), and mortality convergence is observed due to the exhaustion of initial differences in redundancy levels.

Miscellaneous

Recently, a kind of early senescence has been alleged to be a possible unintended outcome of early cloning experiments. The issue was raised in the case of Dolly the sheep, following her death from a contagious lung disease. The claim that Dolly's early death involved premature senescence has been vigorously contested, and Dolly's creator, Dr. Ian Wilmut has expressed the view that her illness and death were probably unrelated to the fact that she was a clone.

A set of rare hereditary (genetic) disorders, each called progeria, has been known for some time. Sufferers exhibit symptoms resembling accelerated aging, including wrinkled skin. The cause of Hutchinson–Gilford progeria syndrome was reported in the journal *Nature* in May 2003. This report suggests that DNA damage, not oxidative stress, is the cause of this form of accelerated aging.

Transcription

Transcription is the process of creating a complementary RNA copy of a sequence of DNA. Both RNA and DNA are nucleic acids, which use base pairs of nucleotides as a complementary language that can be converted back and forth from DNA to RNA by the action of the correct enzymes. During transcription, a DNA sequence is read by RNA polymerase, which produces a complementary, antiparallel RNA strand. As opposed to DNA replication, transcription results in an RNA complement that includes uracil (U) in all instances where thymine (T) would have occurred in a DNA complement.

Transcription can be explained easily in 4 or 5 simple steps, each moving like a wave along the DNA.

1. RNA polymerase unwinds/"unzips" the DNA by breaking the hydrogen bonds between complimentary nucleotides.
2. RNA nucleotides are paired with complementary DNA bases.
3. RNA sugar-phosphate backbone forms with assistance from RNA polymerase.
4. Hydrogen bonds of the untwisted RNA+DNA helix break, freeing the newly synthesized RNA strand.
5. If the cell has a nucleus, the RNA is further processed and then moves through the small nuclear pores to the cytoplasm.

Transcription is the first step leading to gene expression. The stretch of DNA transcribed into an RNA molecule is called a *transcription unit* and encodes at least one gene. If the gene transcribed encodes a protein, the result of transcription is messenger RNA (mRNA), which will then be used to create that protein via the process of translation. Alternatively, the transcribed gene may encode for either ribosomal RNA (rRNA) or transfer RNA (tRNA), other components of the protein-assembly process, or other ribozymes.

A DNA transcription unit encoding for a protein contains not only the sequence that will eventually be directly translated into the protein (the *coding sequence*) but also *regulatory sequences* that direct and regulate the synthesis of that protein. The regulatory sequence before (upstream from) the coding sequence is called the five prime untranslated region (5'UTR), and the sequence following (downstream from) the coding sequence is called the three prime untranslated region (3'UTR).

Transcription has some proofreading mechanisms, but they are fewer and less effective than the controls for copying DNA; therefore, transcription has a lower copying fidelity than DNA replication.

As in DNA replication, DNA is read from 3' → 5' during transcription. Meanwhile, the complementary RNA is created from the 5' '! 3' direction. This means its 5' end is created first in base pairing. Although DNA is arranged as two antiparallel strands in a double helix, only one of the two DNA strands, called the template strand, is used for transcription. This is because RNA is only single-stranded, as opposed to double-stranded DNA. The other DNA strand is called the coding strand, because its sequence is the same as the newly created RNA transcript (except for the substitution of uracil for thymine). The use of only the 3' '! 5' strand eliminates the need for the Okazaki fragments seen in DNA replication.

Transcription is divided into 5 stages: *pre-initiation*, *initiation*, *promoter clearance*, *elongation* and *termination.*

Major Steps

Pre-initiation

In eukaryotes, RNA polymerase, and therefore the initiation of transcription, requires the presence of a core promoter sequence in the DNA. Promoters are regions of DNA that promote transcription and, in eukaryotes, are found at -30, -75, and -90 base pairs upstream from the start site of transcription. Core promoters are sequences within the promoter that are essential for transcription initiation. RNA polymerase is able to bind to core promoters in the presence of various specific transcription factors.

The most common type of core promoter in eukaryotes is a short DNA sequence known as a TATA box, found 25-30 base pairs upstream from the start site of transcription. The TATA box, as a core promoter, is the binding site for a transcription factor known as TATA-binding protein (TBP), which is itself a subunit of another transcription factor, called Transcription Factor II D (TFIID).

After TFIID binds to the TATA box via the TBP, five more transcription factors and RNA polymerase combine around the TATA box in a series of stages to form a preinitiation complex. One transcription factor, DNA helicase, has helicase activity and so is involved in the separating of opposing strands of double-stranded DNA to provide access to a single-stranded DNA template. However, only a low, or basal, rate of transcription is driven by the preinitiation complex alone.

Other proteins known as activators and repressors, along with any associated coactivators or corepressors, are responsible for modulating transcription rate. Thus, preinitiation complex contains: 1. Core Promoter Sequence 2. Transcription Factors 3. DNA Helicase 4. RNA Polymerase 5. Activators and Repressors The transcription preinitiation in archaea is, in essence, homologous to that of eukaryotes, but is much less complex. The archaeal preinitiation complex assembles at a TATA-box binding site; however, in archaea, this complex is composed of only RNA polymerase II, TBP, and TFB (the archaeal homologue of eukaryotic transcription factor II B (TFIIB)).

Initiation

In bacteria, transcription begins with the binding of RNA polymerase to the promoter in DNA. RNA polymerase is a core enzyme consisting of five subunits: 2 α subunits, 1 β subunit, 1 β' subunit, and 1 ω subunit. At the start of initiation, the core enzyme is associated with a sigma factor that aids in finding the appropriate -35 and -10 base pairs downstream of promoter sequences.

Transcription initiation is more complex in eukaryotes. Eukaryotic RNA polymerase does not directly recognize the core promoter sequences. Instead, a collection of proteins called transcription factors mediate the binding of RNA polymerase and the initiation of transcription. Only after certain transcription factors are attached to the promoter does the RNA polymerase bind to it. The completed assembly of transcription factors and RNA polymerase bind to the promoter, forming a transcription initiation complex. Transcription in the archaea domain is similar to transcription in eukaryotes.

Promoter Clearance

After the first bond is synthesized, the RNA polymerase must clear the promoter. During this time there is a tendency to release the RNA transcript and produce truncated transcripts. This is called *abortive initiation* and is common for both eukaryotes and prokaryotes. Abortive initiation continues to occur until the σ factor rearranges, resulting in the transcription elongation complex (which gives a 35 bp moving footprint). The σ factor is released before 80 nucleotides of mRNA are synthesized. Once the transcript reaches approximately 23 nucleotides, it no longer slips and elongation can occur. This, like most of the remainder of transcription, is an energy-dependent process, consuming adenosine triphosphate (ATP).

Promoter clearance coincides with phosphorylation of serine 5 on the carboxy terminal domain of RNA Pol in eukaryotes, which is phosphorylated by TFIIH.

Elongation

One strand of the DNA, the *template strand* (or noncoding strand), is used as a template for RNA synthesis. As transcription proceeds, RNA polymerase traverses the template strand and uses base pairing complementarity with the DNA template to create an RNA copy. Although RNA polymerase traverses the template strand from 3' '! 5', the coding (non-template) strand and newly-formed RNA can also be used as reference points, so transcription can be described as occurring 5' '! 3'. This produces an RNA molecule from 5' '! 3', an exact copy of the coding strand (except that thymines are replaced with uracils, and the nucleotides are composed of a ribose (5-carbon) sugar where DNA has deoxyribose (one less oxygen atom) in its sugar-phosphate backbone).

Unlike DNA replication, mRNA transcription can involve multiple RNA polymerases on a single DNA template and multiple rounds of transcription (amplification of particular mRNA), so many mRNA molecules can be rapidly produced from a single copy of a gene.

Elongation also involves a proofreading mechanism that can replace incorrectly incorporated bases. In eukaryotes, this may correspond with short pauses during transcription that allow appropriate RNA editing factors to bind. These pauses may be intrinsic to the RNA polymerase or due to chromatin structure.

Termination

Bacteria use two different strategies for transcription termination. In Rho-independent transcription termination, RNA transcription stops when the newly synthesized RNA molecule forms a G-C-rich hairpin loop followed by a run of Us. When the hairpin forms, the mechanical stress breaks the weak rU-dA bonds, now filling the DNA-RNA hybrid. This pulls the poly-U transcript out of the active site of the RNA polymerase, in effect, terminating transcription. In the "Rho-dependent" type of termination, a protein factor called "Rho" destabilizes the interaction between the template and the mRNA, thus releasing the newly synthesized mRNA from the elongation complex.

Transcription termination in eukaryotes is less understood but involves cleavage of the new transcript followed by template-independent addition of *A*s at its new 3' end, in a process called polyadenylation.

Measuring and Detecting Transcription

Transcription can be measured and detected in a variety of ways:

- Nuclear Run-on assay: measures the relative abundance of newly formed transcripts
- RNase protection assay and ChIP-Chip of RNAP: detect active transcription sites
- RT-PCR: measures the absolute abundance of total or nuclear RNA levels, which may however differ from transcription rates
- DNA microarrays: measures the relative abundance of the global total or nuclear RNA levels; however, these may differ from transcription rates
- In situ hybridization: detects the presence of a transcript

- MS2 tagging: by incorporating RNA stem loops, such as MS2, into a gene, these become incorporated into newly synthesized RNA. The stem loops can then be detected using a fusion of GFP and the MS2 coat protein, which has a high affinity, sequence-specific interaction with the MS2 stem loops. The recruitment of GFP to the site of transcription is visualised as a single fluorescent spot. This remarkable new approach has revealed that transcription occurs in discontinuous bursts, or pulses. With the notable exception of in situ techniques, most other methods provide cell population averages, and are not capable of detecting this fundamental property of genes.
- Northern blot: the traditional method, and until the advent of RNA-Seq, the most quantitative
- RNA-Seq: applies next-generation sequencing techniques to sequence whole transcriptomes, which allows the measurement of relative abundance of RNA, as well as the detection of additional variations such as fusion genes, post-translational edits and novel splice sites.

Transcription Factories

Active transcription units are clustered in the nucleus, in discrete sites called transcription factories or euchromatin. Such sites can be visualized by allowing engaged polymerases to extend their transcripts in tagged precursors (Br-UTP or Br-U) and immuno-labelling the tagged nascent RNA. Transcription factories can also be localized using fluorescence in situ hybridization or marked by antibodies directed against polymerases. There are ~10,000 factories in the nucleoplasm of a HeLa cell, among which are ~8,000 polymerase II factories and ~2,000 polymerase III factories. Each polymerase II factory contains ~8 polymerases. As most active transcription units are associated with only one polymerase, each factory usually contains ~8 different transcription units. These units might be associated through promoters and/or enhancers, with loops forming a 'cloud' around the factor.

History

A molecule that allows the genetic material to be realized as a protein was first hypothesized by François Jacob and Jacques Monod. RNA synthesis by RNA polymerase was established *in vitro* by several laboratories by 1965; however, the RNA synthesized by these enzymes had properties that suggested the existence of an additional factor needed to terminate transcription correctly. In 1972, Walter Fiers became the first person to actually prove the existence of the terminating enzyme.

Roger D. Kornberg won the 2006 Nobel Prize in Chemistry "for his studies of the molecular basis of eukaryotic transcription".

Reverse Transcription

Some viruses (such as HIV, the cause of AIDS), have the ability to transcribe RNA into DNA. HIV has an RNA genome that is duplicated into DNA. The resulting DNA can be merged with the DNA genome of the host cell. The main enzyme responsible for synthesis of DNA from an RNA template is called reverse transcriptase. In the case of HIV, reverse transcriptase is responsible for synthesizing a complementary DNA strand (cDNA) to the viral RNA genome. An associated enzyme, ribonuclease H, digests the RNA strand, and reverse transcriptase synthesises a complementary strand of DNA to form a double helix DNA structure. This cDNA is integrated into the host cell's genome via another enzyme (integrase) causing the host cell to generate viral proteins that reassemble into new viral particles. In HIV, subsequent to this, the host cell undergoes programmed cell death, apoptosis of T cells. However, in other retroviruses, the host cell remains intact as the virus buds out of the cell.

Some eukaryotic cells contain an enzyme with reverse transcription activity called telomerase. Telomerase is a reverse transcriptase that lengthens the ends of linear chromosomes. Telomerase carries an RNA template from which it synthesizes DNA repeating sequence, or "junk" DNA. This repeated sequence of DNA is important because, every time a linear chromosome is duplicated, it is shortened in length.

With "junk" DNA at the ends of chromosomes, the shortening eliminates some of the non-essential, repeated sequence rather than the protein-encoding DNA sequence farther away from the chromosome end. Telomerase is often activated in cancer cells to enable cancer cells to duplicate their genomes indefinitely without losing important protein-coding DNA sequence. Activation of telomerase could be part of the process that allows cancer cells to become *immortal*. However, the true *in vivo* significance of telomerase has still not been empirically proven.

Ribonucleic Acid

Ribonucleic acid (RNA) is one of the three major macromolecules (along with DNA and proteins) that are essential for all known forms of life.

Like DNA, RNA is made up of a long chain of components called nucleotides. Each nucleotide consists of a nucleobase (sometimes called a nitrogenous base), a ribose sugar, and a phosphate group. The sequence of nucleotides allows RNA to encode genetic information. For example, some viruses use RNA instead of DNA as their genetic material, and all organisms use messenger RNA (mRNA) to carry the genetic information that directs the synthesis of proteins.

Like proteins, some RNA molecules play an active role in cells by catalyzing biological reactions, controlling gene expression, or sensing and communicating responses to cellular signals. One of these active processes is protein synthesis, a universal function whereby mRNA molecules direct the assembly of proteins on ribosomes. This process uses transfer RNA (tRNA) molecules to deliver amino acids to the ribosome, where ribosomal RNA (rRNA) links amino acids together to form proteins.

The chemical structure of RNA is very similar to that of DNA, with two differences—(a) RNA contains the sugar ribose while DNA contains the slightly different sugar deoxyribose (a type of ribose that lacks one oxygen atom), and (b) RNA has the nucleobase uracil while DNA contains thymine (uracil and thymine have similar base-pairing properties). Unlike DNA, most RNA molecules are single-stranded.

Single-stranded RNA molecules adopt very complex three-dimensional structures, since they are not restricted to the repetitive double-helical form of double-stranded DNA. RNA is made within living cells by RNA polymerases, enzymes that act to copy a DNA or RNA template into a new RNA strand through processes known as transcription or RNA replication, respectively.

Comparison with DNA

RNA and DNA are both nucleic acids, but differ in three main ways. First, unlike DNA, which is, in general, double-stranded, RNA is a single-stranded molecule in many of its biological roles and has a much shorter chain of nucleotides. Second, while DNA contains *deoxyribose*, RNA contains *ribose* (in deoxyribose there is no hydroxyl group attached to the pentose ring in the 2' position). These hydroxyl groups make RNA less stable than DNA because it is more prone to hydrolysis. Third, the complementary base to adenine is not thymine, as it is in DNA, but rather uracil, which is an unmethylated form of thymine.

Like DNA, most biologically active RNAs, including mRNA, tRNA, rRNA, snRNAs, and other non-coding RNAs, contain self-complementary sequences that allow parts of the RNA to fold and pair with itself to form double helices. Structural analysis of these RNAs has revealed that they are highly structured. Unlike DNA, their structures do not consist of long double helices but rather collections of short helices packed together into structures akin to proteins.

In this fashion, RNAs can achieve chemical catalysis, like enzymes. For instance, determination of the structure of the ribosome—an enzyme that catalyzes peptide bond formation—revealed that its active site is composed entirely of RNA.

Structure

Each nucleotide in RNA contains a ribose sugar, with carbons numbered 1' through 5'. A base is attached to the 1' position, in general, adenine (A), cytosine (C), guanine (G), or uracil (U). Adenine and guanine are purines, cytosine, and uracil are pyrimidines.

A phosphate group is attached to the 3' position of one ribose and the 5' position of the next. The phosphate groups have a negative charge each at physiological pH, making RNA a charged molecule (polyanion). The bases may form hydrogen bonds between cytosine and guanine, between adenine and uracil and between guanine and uracil. However, other interactions are possible, such as a group of adenine bases binding to each other in a bulge, or the GNRA tetraloop that has a guanine–adenine base-pair.

An important structural feature of RNA that distinguishes it from DNA is the presence of a hydroxyl group at the 2' position of the ribose sugar. The presence of this functional group causes the helix to adopt the A-form geometry rather than the B-form most commonly observed in DNA. This results in a very deep and narrow major groove and a shallow and wide minor groove. A second consequence of the presence of the 2'-hydroxyl group is that in conformationally flexible regions of an RNA molecule (that is, not involved in formation of a double helix), it can chemically attack the adjacent phosphodiester bond to cleave the backbone.

RNA is transcribed with only four bases (adenine, cytosine, guanine and uracil), but these bases and attached sugars can be modified in numerous ways as the RNAs mature. Pseudouridine (Ø), in which the linkage between uracil and ribose is changed from a C–N bond to a C–C bond, and ribothymidine (T) are found in various places (the most notable ones being in the TØC loop of tRNA). Another notable modified base is hypoxanthine, a deaminated adenine base whose nucleoside is called inosine (I). Inosine plays a key role in the wobble hypothesis of the genetic code.

There are nearly 100 other naturally occurring modified nucleosides, of which pseudouridine and nucleosides with 2'-O-methylribose are the most common. The specific roles of many of these modifications in RNA are not fully understood. However, it is notable that, in ribosomal RNA, many of the post-transcriptional modifications occur in highly functional regions, such as the peptidyl transferase center and the subunit interface, implying that they are important for normal function.

The functional form of single stranded RNA molecules, just like proteins, frequently requires a specific tertiary structure. The scaffold for this structure is provided by secondary structural elements that are hydrogen bonds within the molecule. This leads to several recognizable "domains" of secondary structure like hairpin loops, bulges, and internal loops. Since RNA is charged, metal ions such as Mg^{2+} are needed to stabilise many secondary and tertiary structures.

Synthesis

Synthesis of RNA is usually catalyzed by an enzyme—RNA polymerase—using DNA as a template, a process known as transcription. Initiation of transcription begins with the binding of the enzyme to a promoter sequence in the DNA (usually found "upstream" of a gene). The DNA double helix is unwound by the helicase activity of the enzyme. The enzyme then progresses along the template strand in the 3' to 5' direction, synthesizing a complementary RNA molecule with elongation occurring in the 5' to 3' direction. The DNA sequence also dictates where termination of RNA synthesis will occur.

RNAs are often modified by enzymes after transcription. For example, a poly(A) tail and a 5' cap are added to eukaryotic pre-mRNA and introns are removed by the spliceosome. There are also a number of RNA-dependent RNA polymerases that use RNA as their template for synthesis of a new strand of RNA. For instance, a number of RNA viruses (such as poliovirus) use this type of enzyme to replicate their genetic material. Also, RNA-dependent RNA polymerase is part of the RNA interference pathway in many organisms.

Types of RNA

Messenger RNA (mRNA) is the RNA that carries information from DNA to the ribosome, the sites of protein synthesis (translation) in the cell. The coding sequence of the mRNA determines the amino acid sequence in the protein that is produced. Many RNAs do not code for protein however (about 97% of the transcriptional output is non-protein-coding in eukaryotes).

These so-called non-coding RNAs ("ncRNA") can be encoded by their own genes (RNA genes), but can also derive from mRNA introns. The most prominent examples of non-coding RNAs are transfer RNA (tRNA) and ribosomal RNA (rRNA), both of which are involved in the process of translation.

There are also non-coding RNAs involved in gene regulation, RNA processing and other roles. Certain RNAs are able to catalyse chemical reactions such as cutting and ligating other RNA molecules, and the catalysis of peptide bond formation in the ribosome; these are known as ribozymes.

In Translation

Messenger RNA (mRNA) carries information about a protein sequence to the ribosomes, the protein synthesis factories in the cell. It is coded so that every three nucleotides (a codon) correspond to one amino acid. In eukaryotic cells, once precursor mRNA (pre-mRNA) has been transcribed from DNA, it is processed to mature mRNA. This removes its introns—non-coding sections of the pre-mRNA. The mRNA is then exported from the nucleus to the cytoplasm, where it is bound to ribosomes and translated into its corresponding protein form with the help of tRNA. In prokaryotic cells, which do not have nucleus and cytoplasm compartments, mRNA can bind to ribosomes while it is being transcribed from DNA. After a certain amount of time the message degrades into its component nucleotides with the assistance of ribonucleases.

Transfer RNA (tRNA) is a small RNA chain of about 80 nucleotides that transfers a specific amino acid to a growing polypeptide chain at the ribosomal site of protein synthesis during translation. It has sites for amino acid attachment and an anticodon region for codon recognition that binds to a specific sequence on the messenger RNA chain through hydrogen bonding.

Ribosomal RNA (rRNA) is the catalytic component of the ribosomes. Eukaryotic ribosomes contain four different rRNA molecules: 18S, 5.8S, 28S and 5S rRNA. Three of the rRNA molecules are synthesized in the nucleolus, and one is synthesized elsewhere. In the cytoplasm, ribosomal RNA and protein combine to form a nucleoprotein called a ribosome. The ribosome binds mRNA and carries out protein synthesis. Several ribosomes may be attached to a single mRNA at any time. rRNA is extremely abundant and makes up 80% of the 10 mg/ml RNA found in a typical eukaryotic cytoplasm. Transfer-messenger RNA (tmRNA) is found in many bacteria and plastids. It tags proteins encoded by mRNAs that lack stop codons for degradation and prevents the ribosome from stalling.

Regulatory RNAs

Several types of RNA can downregulate gene expression by being complementary to a part of an mRNA or a gene's DNA. MicroRNAs (miRNA; 21-22 nt) are found in eukaryotes and act through RNA interference (RNAi), where an effector complex of miRNA and enzymes can break down mRNA to which the miRNA is complementary, block the mRNA from being translated, or accelerate its degradation. While small interfering RNAs (siRNA; 20-25 nt) are often produced by breakdown of viral RNA, there are also endogenous sources of siRNAs. siRNAs act through RNA interference in a fashion similar to miRNAs. Some miRNAs and siRNAs can cause genes they target to be methylated, thereby decreasing or increasing transcription of those genes. Animals have Piwi-interacting RNAs (piRNA; 29-30 nt) which are active in germline cells and are thought to be a defense against transposons and play a role in gametogenesis.

Many prokaryotes have CRISPR RNAs, a regulatory system similar to RNA interference. Antisense RNAs are widespread; most downregulate a gene, but a few are activators of transcription. One way antisense RNA can act is by binding to an mRNA, forming double-stranded RNA that is enzymatically degraded. There are many long noncoding RNAs that regulate genes in eukaryotes, one such RNA is Xist, which coats one X chromosome in female mammals and inactivates it. An mRNA may contain regulatory elements itself, such as riboswitches, in the 5' untranslated region or 3' untranslated region; these cis-regulatory elements regulate the activity of that mRNA. The untranslated regions can also contain elements that regulate other genes.

In RNA Processing

Many RNAs are involved in modifying other RNAs. Introns are spliced out of pre-mRNA by spliceosomes, which contain several small nuclear RNAs (snRNA), or the introns can be ribozymes that are spliced by themselves. RNA can also be altered by having its nucleotides modified to other nucleotides than A, C, G and U. In eukaryotes, modifications of RNA nucleotides are generally directed by small nucleolar RNAs (snoRNA; 60-300 nt), found in the nucleolus and cajal bodies. snoRNAs associate with enzymes and guide them to a spot on an RNA by basepairing to that RNA. These enzymes then perform the nucleotide modification. rRNAs and tRNAs are extensively modified, but snRNAs and mRNAs can also be the target of base modification.

RNA Genomes

Like DNA, RNA can carry genetic information. RNA viruses have genomes composed of RNA, and a variety of proteins encoded by that genome. The viral genome is replicated by some of those proteins, while other proteins protect the genome as the virus particle moves to a new host cell. Viroids are another group of pathogens, but they consist only of RNA, do not encode any protein and are replicated by a host plant cell's polymerase.

In Reverse Transcription

Reverse transcribing viruses replicate their genomes by reverse transcribing DNA copies from their RNA; these DNA copies are then transcribed to new RNA. Retrotransposons also spread by copying DNA and RNA from one another, and telomerase contains an RNA that is used as template for building the ends of eukaryotic chromosomes.

Double-stranded RNA

Double-stranded RNA (dsRNA) is RNA with two complementary strands, similar to the DNA found in all cells. dsRNA forms the genetic material of some viruses (double-stranded RNA viruses). Double-stranded RNA such as viral RNA or siRNA can trigger RNA interference in eukaryotes, as well as interferon response in vertebrates.

Key Discoveries in RNA Biology

Research on RNA has led to many important biological discoveries and numerous Nobel Prizes. Nucleic acids were discovered in 1868 by Friedrich Miescher, who called the material 'nuclein' since it was found in the nucleus. It was later discovered that prokaryotic cells, which do not have a nucleus, also contain nucleic acids.

The role of RNA in protein synthesis was suspected already in 1939. Severo Ochoa won the 1959 Nobel Prize in Medicine (shared with Arthur Kornberg) after he discovered an enzyme that can synthesize RNA in the laboratory. Ironically, the enzyme discovered by Ochoa (polynucleotide phosphorylase) was later shown to be responsible for RNA degradation, not RNA synthesis.

The sequence of the 77 nucleotides of a yeast tRNA was found by Robert W. Holley in 1965, winning Holley the 1968 Nobel Prize in Medicine (shared with Har Gobind Khorana and Marshall Nirenberg). In 1967, Carl Woese hypothesized that RNA might be catalytic and suggested that the earliest forms of life (self-replicating molecules) could have relied on RNA both to carry genetic information and to catalyze biochemical reactions—an RNA world.

During the early 1970s, retroviruses and reverse transcriptase were discovered, showing for the first time that enzymes could copy RNA into DNA (the opposite of the usual route for transmission of genetic information). For this work, David Baltimore, Renato Dulbecco and Howard Temin were awarded a Nobel Prize in 1975. In 1976, Walter Fiers and his team determined the first complete nucleotide sequence of an RNA virus genome, that of bacteriophage MS2.

In 1977, introns and RNA splicing were discovered in both mammalian viruses and in cellular genes, resulting in a 1993 Nobel to Philip Sharp and Richard Roberts. Catalytic RNA molecules (ribozymes) were discovered in the early 1980s, leading to a 1989 Nobel award to Thomas Cech and Sidney Altman. In 1990 it was found in petunia that introduced genes can silence similar genes of the plant's own, now known to be a result of RNA interference.

At about the same time, 22 nt long RNAs, now called microRNAs, were found to have a role in the development of *C. elegans*. Studies on RNA interference gleaned a Nobel Prize for Andrew Fire and Craig Mello in 2006, and another Nobel was awarded for studies on transcription of RNA to Roger Kornberg in the same year. The discovery of gene regulatory RNAs has led to attempts to develop drugs made of RNA, such as siRNA, to silence genes.

Transcription and Processing of RNA

Gene Expression: Transcription

The majority of genes are expressed as the proteins they encode. The process occurs in two steps:

- Transcription = DNA '! RNA
- Translation = RNA '! protein.

Taken together, they make up the "central dogma" of biology: DNA '! RNA '! protein.

Gene Transcription: DNA '! RNA

DNA serves as the template for the synthesis of RNA much as it does for its own replication.

The Steps

- Some 50 different protein transcription factors bind to promoter sites, usually on the 52 side of the gene to be transcribed.
- An enzyme, an RNA polymerase, binds to the complex of transcription factors.
- Working together, they open the DNA double helix.
- The RNA polymerase proceeds to "read" one strand moving in its 32 '! 52 direction.
- In eukaryotes, this requires — at least for protein-encoding genes — that the nucleosomes in front of the advancing RNA polymerase (RNAP II) be removed. A complex of proteins is responsible for this. The same complex replaces the nucleosomes after the DNA has been transcribed and RNAP II has moved on.
- As the RNA polymerase travels along the DNA strand, it assembles ribonucleotides (supplied as triphosphates, e.g., ATP) into a strand of RNA.
- Each ribonucleotide is inserted into the growing RNA strand following the rules of base pairing. Thus for each C encountered on the DNA strand, a G is inserted in the RNA; for each G, a C; and for each T, an A. However, each A on the DNA guides the insertion of the pyrimidine uracil (U, from uridine triphosphate, UTP). There is no T in RNA.

Quality control: Occasionally RNA polymerase will select and insert an incorrect, mismatched, ribonucleotide. When this occurs in bacteria (and perhaps in all organisms), the enzyme backs up, removes the incorrect nucleotide (and the one preceding it) and tries again. (Described by Zenkin *et al.*, in the 28 July 2006 issue of Science.)

- Synthesis of the RNA proceeds in the 52 '! 32 direction.
- As each nucleoside triphosphate is brought in to add to the 32 end of the growing strand, the two terminal phosphates are removed.
- When transcription is complete, the transcript is released from the polymerase and, shortly thereafter, the polymerase is released from the DNA.

Note that at any place in a DNA molecule, either strand may be serving as the template; that is, some genes "run" one way, some the other (and in a few remarkable cases, the same segment of double helix contains genetic information on both strands!). In all cases, however, RNA polymerase transcribes the DNA strand in its 32 '! 52 direction.

A report in the 4 January 2001 issue of Nature shows that RNA polymerase actually tracks around the double helix of DNA. In vitro, at least, when RNA polymerase is immobilized, it spins the DNA molecule around and around as it moves along the molecule. Whether it is the polymerase or the DNA that does the spinning in vivo remains to be determined.

Types of RNA

Sedimentation pattern produced by high-speed centrifugation of RNA extracted from the precursors of rabbit red blood cells. The discrete bands represent particular classes of RNA.

The transfer RNAs band at about 4S. The ribosomal RNAs of eukaryotes sediment at 5S, 5.8S, 18S, and 28S. (The larger the sedimentation unit, S, the larger the molecule — but not proportionally.) The RNA forming the band at 9S is the messenger RNA for the synthesis of hemoglobin, the major protein synthesized in these cells. In most types of cells, the messenger RNAs are extremely heterogenous, with small amounts distributed from 6S to 25S.

Several types of RNA are synthesized in the nucleus of eukaryotic cells. Of particular interest are:

- messenger RNA (mRNA). This will later be translated into a polypeptide.
- ribosomal RNA (rRNA). This will be used in the building of ribosomes: machinery for synthesizing proteins by translating mRNA.
- transfer RNA (tRNA). RNA molecules that carry amino acids to the growing polypeptide.
- small nuclear RNA (snRNA). DNA transcription of the genes for mRNA, rRNA, and tRNA produces large precursor molecules ("primary

transcripts") that must be processed within the nucleus to produce the functional molecules for export to the cytosol. Some of these processing steps are mediated by snRNAs.

- small nucleolar RNA (snoRNA). These RNAs within the nucleolus have several functions (described below).
- microRNA (miRNA). These are tiny (~22 nucleotides) RNA molecules that appear to regulate the expression of messenger RNA (mRNA) molecules. [Discussion]
- XIST RNA. This inactivates one of the two X chromosomes in female vertebrates. [Discussion].

Messenger RNA (mRNA)

Messenger RNA comes in a wide range of sizes reflecting the size of the polypeptide it encodes. Most cells produce small amounts of thousands of different mRNA molecules, each to be translated into a peptide needed by the cell.

Many mRNAs are common to most cells, encoding "housekeeping" proteins needed by all cells (e.g. the enzymes of glycolysis). Other mRNAs are specific for only certain types of cells. These encode proteins needed for the function of that particular cell (e.g., the mRNA for hemoglobin in the precursors of red blood cells).

Ribosomal RNA (rRNA)

There are 4 kinds. In eukaryotes, these are:

- 18S rRNA. One of these molecules, along with some 30 different protein molecules, is used to make the small subunit of the ribosome.
- 28S, 5.8S, and 5S rRNA. One each of these molecules, along with some 45 different proteins, are used to make the large subunit of the ribosome.

The S number given each type of rRNA reflects the rate at which the molecules sediment in the ultracentrifuge. The larger the number, the larger the molecule (but not proportionally).

The 28S, 18S, and 5.8S molecules are produced by the processing of a single primary transcript from a cluster of identical copies of a single gene. The 5S molecules are produced from a different cluster of identical genes.

Transfer RNA (tRNA)

There are some 32 different kinds of tRNA in a typical eukaryotic cell.

- Each is the product of a separate gene.
- They are small (~4S), containing 73-93 nucleotides.

- Many of the bases in the chain pair with each other forming sections of double helix.
- The unpaired regions form 3 loops.
- Each kind of tRNA carries (at its 32 end) one of the 20 amino acids (thus most amino acids have more than one tRNA responsible for them).
- At one loop, 3 unpaired bases form an anticodon.
- Base pairing between the anticodon and the complementary codon on a mRNA molecule brings the correct amino acid into the growing polypeptide chain.

Small Nuclear RNA (snRNA)

Approximately a dozen different genes for snRNAs, each present in multiple copies, have been identified. The snRNAs have various roles in the processing of the other classes of RNA.

For example, several snRNAs are part of the spliceosomes that participate in converting pre-mRNA into mRNA by excising the introns and splicing the exons. [Link down to the discussion of RNA processing.]

Small Nucleolar RNA (snoRNA)

As the name suggests, these small (60–300 nucleotides) RNAs are found in the nucleolus where they are responsible for several functions:

- Some participate in making ribosomes by helping to cut up the large RNA precursor of the 28S, 18S, and 5.8S molecules.
- Others chemically modify many of the nucleotides in rRNA, tRNA, and snRNA molecules, e.g., by adding methyl groups to ribose.
- Some have been implicated in the alternative splicing of pre-mRNA to different forms of mature mRNA.
- One snoRNA serves as the template for the synthesis of telomeres.

In vertebrates, the snoRNAs are made from introns removed during RNA processing.

Noncoding RNA

Only messenger RNA encodes polypeptides. All the other classes of RNA, including types not mentioned here, are thus called noncoding RNA. Much remains to be learned about the function(s) of some of them. But, taken together, noncoding RNAs probably account for two-thirds of the transcription going on in the nucleus.

The RNA Polymerases

The RNA polymerases are huge multi-subunit protein complexes. Three kinds are found in eukaryotes.

- RNA polymerase I (Pol I). It transcribes the rRNA genes for the precursor of the 28S, 18S, and 5.8S molecules (and is the busiest of the RNA polymerases).
- RNA polymerase II (Pol II; also known as RNAP II). It transcribes protein-encoding genes into mRNA (and also the snRNA genes).
- RNA polymerase III (Pol III). It transcribes the 5S rRNA genes and all the tRNA genes.

Two remarkable reports in the 8 June 2001 issue of Science show the structure of Pol II and reveal many details of how it uses the antisense strand of DNA to synthesize a strand of mRNA.

RNA Processing: Pre-mRNA '! mRNA

All the primary transcripts produced in the nucleus must undergo processing steps to produce functional RNA molecules for export to the cytosol. We shall confine ourselves to a view of the steps as they occur in the processing of pre-mRNA to mRNA. Most eukaryotic genes are split into segments. In decoding the open reading frame of a gene for a known protein, one usually encounters periodic stretches of DNA calling for amino acids that do not occur in the actual protein product of that gene. Such stretches of DNA, which get transcribed into RNA but not translated into protein, are called introns. Those stretches of DNA that do code for amino acids in the protein are called exons. Examples:

- The gene for one type of collagen found in chickens is split into 52 separate exons.
- The gene for dystrophin, which is mutated in boys with muscular dystrophy, has 79 exons.
- Even the genes for rRNA and tRNA are split by introns.
- The human genome is estimated to contain some 180,000 exons. With a current estimate of 21,000 genes, the average exon content of our genes is about 9.

In general, introns tend to be much longer than exons. An average eukaryotic exon is only 140 nucleotides long, but one human intron stretches for 480,000 nucleotides!

Removal of the introns — and splicing the exons together — are among the essential steps in synthesizing mRNA.

The Steps of RNA Processing

- Synthesis of the cap. This is a modified guanine (G) which is attached to the 52 end of the pre-mRNA as it emerges from RNA polymerase II (RNAP II). The cap
  - — protects the RNA from being degraded by enzymes that degrade RNA from the 52 end;
  - — serves as an assembly point for the proteins needed to recruit the small subunit of the ribosome to begin translation.
- Step-by-step removal of introns present in the pre-mRNA and splicing of the remaining exons. This step takes place as the pre-mRNA continues to emerge from RNAP II.
- Synthesis of the poly(A) tail. This is a stretch of adenine (A) nucleotides. When a special poly(A) attachment site in the pre-mRNA emerges from RNAP II, the transcript is cut there, and the poly(A) tail is attached to the exposed 32 end. This completes the mRNA molecule, which is now ready for export to the cytosol. (The remainder of the transcript is degraded, and the RNA polymerase leaves the DNA).

The cutting and splicing of mRNA must be done with great precision. If even one nucleotide is left over from an intron or one is removed from an exon, the reading frame from that point on will be shifted, producing new codons specifying a totally different sequence of amino acids from that point to the end of the molecule (which often ends prematurely anyway when the shifted reading frame generates a STOP codon).

The removal of introns and splicing of exons is done by spliceosomes. These are a complexes of 5 snRNA molecules and some 145 different proteins. The introns in most pre-mRNAs begin with a GU and end with an AG. Presumably these short sequences assist in guiding the spliceosome.

Visual Evidence

The upper image is an electron micrograph of a mRNA-DNA hybrid molecule formed by mixing the messenger RNA (mRNA) from a clone of antibody-secreting cells with single-stranded DNA from the same kind of cells. The bar represents the length of 1000 bases.

The lower diagram is an interpretation of the micrograph. The solid line represents the DNA; the dotted line the mRNA. The loops (I_A, I_B, etc.) represent the introns that separate the exons encoding the domains of an antibody heavy chain:

- V = variable region
- E_1 = first constant region (C_H1) domain

- E_H = hinge region
- E_2 and E_3 = the nucleotides encoding the two C-terminal domains (C_H2 and C_H3).

The unhybridized portion of the mRNA is its poly(A) tail.

Alternative Splicing

The processing of pre-mRNA for many proteins proceeds along various paths in different cells or under different conditions. For example, early in the differentiation of a B cell (a lymphocyte that synthesizes an antibody) the cell first uses an exon that encodes a transmembrane domain that causes the molecule to be retained at the cell surface. Later, the B cell switches to using a different exon whose domain enables the protein to be secreted from the cell as a circulating antibody molecule.

Alternative splicing provides a mechanism for producing a wide variety of proteins from a small number of genes. While we humans may turn out to have only some 20 thousand genes, we probably make at least 10 times that number of different proteins. It is now estimated that 92–94% of our genes produce pre-mRNAs that are alternatively-spliced. There is evidence that the pattern of alternative splicing differs consistently in different tissues and so must be regulated. But whether all the products are functional or that many are simply the outcome of an error-prone process remains to be seen.

Alternative splicing not only provides different proteins from a single gene but also different 3' UTRs and 5' UTRs. Although not translated into protein, these untranslated regions contain signals that, for example, dictate where in the cell that protein will accumulate. Two examples:

- The 3' UTR of the *bicoid* gene in Drosophila directs the mRNA to the anterior of the embryo;
- the same region in the *VegT* gene of Xenopus directs its mRNA to the vegetal pole of the embryo.

One of the most dramatic examples of alternative splicing is the *Dscam* gene in Drosophila. This single gene contains some 116 exons of which 17 are retained in the final mRNA. Some exons are always included; others are selected from an array. Theoretically this system is able to produce 38,016 different proteins. And, in fact, over 18,000 different ones have been found in Drosophila hemolymph.

These Dscam proteins are used to establish a unique identity for each neuron. It works like this. Each developing neuron draws upon the pool of thousands of possible different mRNAs to synthesize a dozen or so of them. Which ones are selected appears to be simply a matter of chance, but because of the size of the pool, each neuron will most likely end up with a unique set of a dozen or so Dscam proteins. As each developing neuron in the central nervous system sprouts dendrites and an axon, these express its unique collection of Dscam proteins.

If the extensions of a single neuron should meet each other in the tangled web that is the hallmark of nervous tissue, they are repelled. In this way, thousands of different neurons can coexist in intimate contact without the danger of nonfunctional contacts between the various extensions of the same neuron.

We also have a *DSCAM* gene which is located on chromosome 21. When present in 3 copies, it is probably responsible for some of the features of Down syndrome (and accounts for its full name: Down Syndrome Cell Adhesion Molecule). Perhaps the incredible diversity of synaptic junctions in the mammalian c.n.s. ($\sim 10^{14}$) is mediated by alternative splicing of a limited number of gene transcripts.

So, whether a particular segment of RNA will be retained as an exon or excised as an intron can vary under different circumstances, such as;

- what type of cell the gene is in;
- what stage of differentiation that cell is passing through;
- what extracellular signals that cell is receiving.

Clearly the switching to an alternate splicing pathway must be closely regulated.

Trans-splicing

Most genes are transcribed and their transcripts processed as described above. RNA polymerase travels down a single strand of a singe gene locus to form pre-mRNA that is processed (including removal of introns) to form the mature mRNA. But there are exceptions. A number of cases have been found where two different precursor transcripts have been spliced together to form the final RNA molecule. The phenomenon is called *trans*-splicing.

Examples: synthesis of a single RNA molecule by splicing together transcripts from ;

- loci located far apart on the same chromosome;
- opposite strands of the same gene locus;
- the two alleles of the gene on their separate (homologous) chromosomes.

The biological importance of these *trans*-spliced transcripts is still unknown for most of them.

Why Split Genes?

Perhaps during evolution, eukaryotic genes have been assembled from smaller, primitive genes - today's exons. Some proteins, like the antibodies mentioned in the previous section, are organized in a set of separate sections or domains each with a special function to perform in the complete molecule. Each domain is encoded by a separate exon. Having the different functional parts of the antibody molecule encoded by separate exons makes it possible to use these units in different combinations. Thus a set of exons in the genome may be the genetic equivalent of the various modular pieces in a box of "Lego" for children to assemble in whatever forms they wish. But the boundaries of other exons do not seem to correspond to domain boundaries of the protein. So perhaps the major benefit of split genes is simply the opportunity they provide for making many different proteins from a single gene through alternative splicing.

Summary

Gene expression occurs in two steps:

- transcription of the information encoded in DNA into a molecule of RNA (described here) and
- translation of the information encoded in the nucleotides of mRNA into a defined sequence of amino acids in a protein (discussed in Gene Translation: RNA → Protein).

Messenger RNA

Messenger RNA (mRNA) is a molecule of RNA encoding a chemical "blueprint" for a protein product. mRNA is transcribed from a DNA template, and carries coding information to the sites of protein synthesis: the ribosomes. Here, the nucleic acid polymer is translated into a polymer of amino acids: a protein. In mRNA as in DNA, genetic information is encoded in the sequence of nucleotides arranged into codons consisting of three bases each. Each codon encodes for a specific amino acid, except the stop codons that terminate protein synthesis. This process requires two other types of RNA: transfer RNA (tRNA) mediates recognition of the codon and provides the corresponding amino acid, while ribosomal RNA (rRNA) is the central component of the ribosome's protein manufacturing machinery.

Synthesis, Processing, and Function

The brief existence of an mRNA molecule begins with transcription and ultimately ends in degradation. During its life, an mRNA molecule may also be processed, edited, and transported prior to translation. Eukaryotic mRNA molecules often require extensive processing and transport, while prokaryotic molecules do not.

Transcription

During transcription, RNA polymerase makes a copy of a gene from the DNA to mRNA as needed. This process is similar in eukaryotes and prokaryotes. One notable difference, however, is that prokaryotic RNA polymerase associates with mRNA processing enzymes during transcription so that processing can proceed quickly after the start of transcription. The short-lived, unprocessed or partially processed, product is termed *pre-mRNA*; once completely processed, it is termed *mature mRNA*.

Eukaryotic Pre-mRNA Processing

Processing of mRNA differs greatly among eukaryotes, bacteria and archea. Non-eukaryotic mRNA is essentially mature upon transcription and requires no processing, except in rare cases. Eukaryotic pre-mRNA, however, requires extensive processing.

5' Cap Addition

A *5' cap* (also termed an RNA cap, an RNA 7-methylguanosine cap or an RNA m^7G cap) is a modified guanine nucleotide that has been added to the "front" or 5' end of a eukaryotic messenger RNA shortly after the start of transcription. The 5' cap consists of a terminal 7-methylguanosine residue which is linked through a 5'-5'-triphosphate bond to the first transcribed nucleotide. Its presence is critical for recognition by the ribosome and protection from RNases. Cap addition is coupled to transcription, and occurs co-transcriptionally, such that each influences the other. Shortly after the start of transcription, the 5' end of the mRNA being synthesized is bound by a cap-synthesizing complex associated with RNA polymerase. This enzymatic complex catalyzes the chemical reactions that are required for mRNA capping. Synthesis proceeds as a multi-step biochemical reaction.

Splicing

Splicing is the process by which pre-mRNA is modified to remove certain stretches of non-coding sequences called introns; the stretches that remain include protein-coding sequences and are called exons. Sometimes pre-mRNA messages may be spliced in several different ways, allowing a single gene to encode multiple proteins. This process is called alternative splicing. Splicing is usually performed by an RNA-protein complex called the spliceosome, but some RNA molecules are also capable of catalyzing their own splicing.

Ribozyme

A ribozyme (from ribonucleic acid enzyme, also called RNA enzyme or catalytic RNA) is an RNA molecule possessing a well defined tertiary structure that enables it to catalyze a chemical reaction. Many natural ribozymes catalyze either the hydrolysis of one of their own phosphodiester bonds (self-cleaving ribozymes), or the hydrolysis of bonds in other RNAs, but they have also been found to catalyze the aminotransferase activity of the ribosome.

Investigators studying the origin of life have produced ribozymes in the laboratory that are capable of catalyzing their own synthesis under very specific conditions, such as an RNA polymerase ribozyme. Mutagenesis and selection has been performed resulting in isolation of improved variants of the "Round-18" polymerase ribozyme from 2001. "B6.61" is able to add up to 20 nucleotides to a primer template in 24 hours, until it decomposes by hydrolysis of its phosphodiester bonds. Some ribozymes may play an important role as therapeutic agents, as enzymes which tailor defined RNA sequences, as biosensors, and for applications in functional genomics and gene discovery.

Discovery

Before the discovery of ribozymes, enzymes, which are defined as catalytic proteins, were the only known biological catalysts. In 1967, Carl Woese, Francis Crick, and Leslie Orgel were the first to suggest that RNA could act as a catalyst. This idea was based upon the discovery that RNA can form complex secondary structures. The first ribozymes were discovered in the 1980s by Thomas R. Cech, who was studying RNA splicing in the ciliated protozoan *Tetrahymena thermophila* and Sidney Altman, who was working on the bacterial RNase P complex.

These ribozymes were found in the intron of an RNA transcript, which removed itself from the transcript, as well as in the RNA component of the RNase P complex, which is involved in the maturation of pre-tRNAs. In 1989, Thomas R. Cech and Sidney Altman won the Nobel Prize in chemistry for their "discovery of catalytic properties of RNA." The term *ribozyme* was first introduced by Kelly Kruger *et al.* in 1982 in a paper published in *Cell.*

It had been a firmly established belief in biology that catalysis was reserved for proteins. In retrospect, catalytic RNA makes a lot of sense. This is based on the old question regarding the origin of life: Which comes first, enzymes that do the work of the cell or nucleic acids that carry the information required to produce the enzymes? Nucleic acids as catalysts circumvents this problem.

In the 1970s Thomas Cech, at the University of Colorado at Boulder, was studying the excision of introns in a ribosomal RNA gene in *Tetrahymena thermophila.* While trying to purify the enzyme responsible for splicing reaction, he found that intron could be spliced out in the absence of any added cell extract. As much as they tried, Cech and his colleagues could not identify any protein associated with the splicing reaction. After much work, Cech proposed that the intron sequence portion of the RNA could break and reform phosphodiester bonds. At about the same time, Sidney Altman, who is a Professor at Yale University, was studying the way tRNA molecules are processed in the cell when he and his colleagues isolated an enzyme called RNase-P, which is responsible for conversion of a precursor tRNA into the active tRNA. Much to their surprise, they found that RNase-P contained RNA in addition to protein and that RNA was an essential component of the active enzyme. This was such a foreign idea that they had difficulty publishing their findings. The following year, Altman demonstrated that RNA can act as a catalyst by showing that the RNase-P RNA subunit could catalyze the cleavage of precursor tRNA into active tRNA in the absence of any protein component.

Since Cech's and Altman's discovery, other investigators have discovered other examples of self-cleaving RNA or catalytic RNA molecules. Many ribozymes have either a hairpin – or hammerhead – shaped active center and a unique secondary structure that allows them to cleave other RNA molecules at specific sequences. It is now possible to make ribozymes that will specifically cleave any RNA molecule. These RNA catalysts may have pharmaceutical applications. For example, a ribozyme has been designed to cleave the RNA of HIV. If such a ribozyme was made by a cell, all incoming virus particles would have their RNA genome cleaved by the ribozyme, which would prevent infection.

Activity

Although most ribozymes are quite rare in the cell, their roles are sometimes essential to life. For example, the functional part of the ribosome, the molecular machine that translates RNA into proteins, is fundamentally a ribozyme, composed of RNA tertiary structural motifs that are often coordinated to metal ions such as Mg^{2+} as cofactors. There is no requirement for divalent cations in a five-nucleotide RNA that can catalyze *trans*-phenylalanation of a four-nucleotide substrate which has three base complementary sequence with the catalyst. The catalyst and substrate were devised by truncation of the C3 ribozyme. RNA can also act as a hereditary molecule, which encouraged Walter Gilbert to propose that in the distant past, the cell used RNA as both the genetic material and the structural and catalytic molecule, rather than dividing these functions between DNA and protein as they are today. This hypothesis became known as the "RNA world hypothesis" of the origin of life.

If ribozymes were the first molecular machines used by early life, then today's remaining ribozymes — such as the ribosome machinery — could be considered living fossils of a life based primarily on nucleic acids. A recent test-tube study of prion folding suggests that an RNA may catalyze the pathological protein conformation in the manner of a chaperone enzyme.

Known Ribozymes

Naturally occurring ribozymes include:

- Peptidyl transferase 23S rRNA
- RNase P
- Group I and Group II introns
- GIR1 branching ribozyme
- Leadzyme - Although initially created *in vitro*, natural examples have been found
- Hairpin ribozyme
- Hammerhead ribozyme
- HDV ribozyme
- Mammalian CPEB3 ribozyme
- VS ribozyme
- *glmS* ribozyme
- CoTC ribozyme.

Artificial Ribozymes

Since the discovery of ribozymes that exist in living organisms, there has been interest in the study of new synthetic ribozymes made in the laboratory. For example, artificially-produced self-cleaving RNAs that have good enzymatic activity have been produced. Tang and Breaker isolated self-cleaving RNAs by in vitro selection of RNAs originating from random-sequence RNAs. Some of the synthetic ribozymes that were produced had novel structures, while some were similar to the naturally occurring hammerhead ribozyme. The techniques used to discover artificial ribozymes involve Darwinian evolution. This approach takes advantage of RNA's dual nature as both a catalyst and an informational polymer, making it easy for an investigator to produce vast populations of RNA catalysts using polymerase enzymes. The ribozymes are mutated by reverse transcribing them with reverse transcriptase into various cDNA and amplified with mutagenic PCR. The selection parameters in these experiments often differ. One approach for selecting a ligase ribozyme involves using biotin tags, which are covalently linked to the substrate. If a molecule possesses the desired ligase activity, a streptavidin matrix can be used to recover the active molecules. Lincoln and Joyce developed an RNA enzyme system capable of self replication in about an hour. By utilizing molecular competition (*in vitro* evolution) of a candidate enzyme mixture, a pair of RNA enzymes emerged, in which each synthesizes the other from synthetic oligonucleotides, with no protein present.

Applications

A type of synthetic ribozyme directed against HIV RNA called gene shears has been developed and has entered clinical testing for HIV infection.

Editing

In some instances, an mRNA will be edited, changing the nucleotide composition of that mRNA. An example in humans is the apolipoprotein B mRNA, which is edited in some tissues, but not others. The editing creates an early stop codon, which upon translation, produces a shorter protein.

Polyadenylation

Polyadenylation is the addition of a poly(A) tail to an RNA molecule. The poly(A) tail consists of multiple adenosine monophosphates; in other words, it is a stretch of RNA which only has adenine bases. In eukaryotes, polyadenylation is part of the process that produces mature messenger RNA (mRNA) for translation. It therefore forms part of the larger process of gene expression.

The process of polyadenylation begins as the transcription of a gene finishes. The 3'-most segment of the newly-made RNA is first cleaved off by a set of proteins; these proteins then synthesize the poly(A) tail at the RNA's 3' end. In some genes these proteins may add a poly(A) tail at any one of several possible sites, polyadenylation can therefore produce more than one transcript from a single gene, similar to alternative splicing.

The poly(A) tail is important for the nuclear export, translation and stability of mRNA. The tail is shortened over time and when it is short enough, the mRNA is enzymatically degraded. However, in a few cell types, mRNAs with short poly(A) tails are stored for later activation by re-polyadenylation in the cytosol. In contrast, when polyadenylation occurs in bacteria, it promotes RNA degradation. This is also sometimes the case for eukaryotic non-coding RNAs. The wide distribution of polyadenylation among living organisms indicates that this process evolved early in the history of life on Earth.

Primer on RNA

RNAs are a type of large biological molecules, whose individual building blocks are called nucleotides. The name *poly(A) tail* (for polyadenylic acid tail) reflects the way RNA nucleotides are abbreviated, with a letter for the base the nucleotide contains (A for adenine, C for cytosine, G for guanine and U for uracil). RNAs are produced (*transcribed*) from a DNA template. By convention, RNA sequences are written in a 5' to 3' direction. The 5' end is the part of the RNA molecule that is transcribed first, and the 3' end is transcribed last. The 3' end is also where the poly(A) tail is found on polyadenylated RNAs.

Messenger RNA (mRNA) is RNA that has a coding region that acts as a template for protein synthesis (*translation*). The rest of the mRNA, the *untranslated regions*, tune how active the mRNA is. There are also many RNAs that are not translated, called non-coding RNAs. Like the untranslated regions, many of these non-coding RNAs have regulatory roles.

Nuclear Polyadenylation

Function

In nuclear polyadenylation, a poly(A) tail is added to an RNA at the end of transcription. On mRNAs, the poly(A) tail protects the mRNA molecule from enzymatic degradation in the cytoplasm and aids in transcription termination, export of the mRNA from the nucleus, and translation. Almost all eukaryotic mRNAs are polyadenylated, with the exception of animal replication-dependent histone mRNAs. These are the only mRNAs in eukaryotes that lack a poly(A) tail, ending instead in a stem-loop structure followed by a purine-rich sequence, termed histone downstream element, that directs where the RNA is cut so that the 3' end of the histone mRNA is formed.

Many eukaryotic non-coding RNAs are always polyadenylated at the end of transcription. There are small RNAs where the poly(A) tail is only seen in intermediary forms and not in the mature RNA as the ends are removed during processing, notably microRNAs. But for many long noncoding RNAs – a seemingly large group of regulatory RNAs that for example includes the RNA Xist which mediates X chromosome inactivation – a poly(A) tail is part of the mature RNA.

Mechanism

The polyadenylation machinery in the nucleus of eukaryotes works on products of RNA polymerase II, such as precursor mRNA. Here, a multi-protein complex cleaves the 3'-most part of a newly produced RNA and polyadenylates the end produced by this cleavage. The cleavage is catalysed by the enzyme CPSF and occurs 10–30 nucleotides downstream of its binding site.

This site is often the sequence AAUAAA on the RNA, but variants of it exist that bind more weakly to CPSF. Two other proteins add specificity to the binding to an RNA: CstF and CFI. CstF binds to a GU-rich region further downstream of CPSF's site. CFI recognises a third site on the RNA (a set of UGUAA sequences in mammals) and can recruit CPSF even if the AAUAAA sequence is missing. The polyadenylation signal – the sequence motif recognised by the RNA cleavage complex – varies between groups of eukaryotes. Most human polyadenylation sites contain the AAUAAA sequence, but this sequence is less common in plants and fungi.

The RNA is cleaved right after transcription, as CstF also binds to RNA polymerase II. Cleavage also involves the protein CFII, though it is unknown how. The cleavage site associated with a polyadenylation signal can vary up to some 50 nucleotides. When the RNA is cleaved, polyadenylation starts, catalysed by polyadenylate polymerase. Polyadenylate polymerase builds the poly(A) tail by adding adenosine monophosphate units from adenosine triphosphate to the RNA, cleaving off pyrophosphate. Another protein, PAB2, binds to the new, short poly(A) tail and increases the affinity of polyadenylate polymerase for the RNA. When the poly(A) tail is approximately 250 nucleotides long the enzyme can no longer bind to CPSF and polyadenylation stops, thus determining the length of the poly(A) tail. CPSF is in contact with RNA polymerase II, allowing it to tell the polymerase to terminate transcription. The polyadenylation machinery is also physically linked to the spliceosome, a complex that removes introns from RNAs.

Downstream Effects

The poly(A) tail acts as the binding site for poly(A)-binding protein. Poly(A)-binding protein promotes export from the nucleus and translation, and inhibits degradation. This protein binds to the poly(A) tail prior to mRNA export from the nucleus and in yeast also recruits poly(A) nuclease, an enzyme that shortens the poly(A) tail and allows the export of the mRNA. Poly(A)-binding protein is exported to the cytoplasm with the RNA. mRNAs that are not exported are degraded by the exosome. Poly(A)-binding protein also can bind to, and thus recruit, several proteins that affect translation, one of these is initiation factor-4G which in turn recruits the 40S ribosomal subunit. However, a poly(A) tail is not required for the translation of all mRNAs.

Deadenylation

In eukaryotic somatic cells, the poly(A) tail of most mRNAs in the cytoplasm gradually get shorter, and mRNAs with shorter poly(A) tail are translated less and degraded sooner. However, it can take many hours before an mRNA is degraded. This deadenylation and degradation process can be accelerated by microRNAs complementary to the 3' untranslated region of an mRNA. In immature egg cells, mRNAs with shortened poly(A) tails are not degraded, but are instead stored without being translated. They are then activated by cytoplasmic polyadenylation after fertilisation, during egg activation.

In animals, poly(A) ribonuclease (PARN) can bind to the 5' cap and remove nucleotides from the poly(A) tail. The level of access to the 5' cap and poly(A) tail is important in controlling how soon the mRNA is degraded. PARN deadenylates less if the RNA is bound by the initiation factors 4E (at the 5' cap) and 4G (at the poly(A) tail), so this is why translation reduces deadenylation. The rate of deadenylation may also be regulated by RNA-binding proteins. Once the poly(A) tail is removed, the decapping complex removes the 5' cap, leading to a degradation of the RNA. Several other enzymes that seem to be involved in deadenylation have been identified in yeast.

Alternative Polyadenylation

Many protein-coding genes have more than one polyadenylation site, so a gene can code for several mRNAs that differ in their 3' end. Since alternative polyadenylation changes the length of the 3' untranslated region, it can change which binding sites for microRNAs the 3' untranslated region contains. MicroRNAs tend to repress translation and promote degradation of the mRNAs they bind to, although there are examples of microRNAs that stabilise transcripts. Alternative polyadenylation can also shorten the coding region, thus making the mRNA code for a different protein, but this is much less common than just shortening the 3' untranslated region.

The choice of poly(A) site can be influenced by extracellular stimuli and depends on the expression of the proteins that take part in polyadenylation. For example, the expression of CstF-64, a subunit of cleavage stimulatory factor (CstF), increases in macrophages in response to lipopolysaccharides (a group of bacterial compounds that trigger an immune response). This results in the selection of weak poly(A) sites and thus shorter transcripts. This removes regulatory elements in the 3' untranslated regions of mRNAs for defense-related products like lysozyme and TNF-α. These mRNAs then have longer half-lives and produce more of these proteins. RNA-binding proteins other than those in the polyadenylation machinery can also affect whether a polyadenyation site is used,as can DNA methylation near the polyadenylation signal.

Cytoplasmic Polyadenylation

There is polyadenylation in the cytosol of some animal cell types, namely in the germ line, during early embryogenesis and in post-synaptic sites of nerve cells. This lengthens the poly(A) tail of an mRNA with a shortened poly(A) tail, so that the mRNA will be translated. These shortened poly(A) tails are often less than 20 nucleotides, and are lengthened to around 80–150 nucleotides.

In the early mouse embryo, cytoplasmic polyadenylation of maternal RNAs from the egg cell allows the cell to survive and grow even though transcription does not start until the middle of the 2-cell stage (4-cell stage in human). In the brain, cytoplasmic polyadenylation is active during learning and could play a role in long-term potentiation, which is the strengthening of the signal transmission from a nerve cell to another in response to nerve impulses and is important for learning and memory formation. Cytoplasmic polyadenylation requires the RNA-binding proteins CPSF and CPEB, and can involve other RNA-binding proteins like Pumilio. Depending on the cell type, the polymerase can be the same type of polyadenylate polymerase (PAP) that is used in the nuclear process, or the cytoplasmic polymerase GLD-2.

Tagging for Degradation in Eukaryotes

For many non-coding RNAs, including tRNA, rRNA, snRNA and snoRNA, polyadenylation is a way of marking the RNA for degradation, in at least yeast. This polyadenylation is done in the nucleus by the TRAMP complex, which adds a tail that is around 40 nucleotides long to the 3' end. The RNA is then degraded by the exosome. Poly(A) tails have also been found on human rRNA fragments, both the form of homopolymeric (A only) and heterpolymeric (mostly A) tails.

In Prokaryotes and Organelles

In many bacteria, both mRNAs and non-coding RNAs can be polyadenylated. This poly(A) tail promotes degradation by the degradosome, which contains two RNA-degrading enzymes: polynucleotide phosphorylase and RNase E. Polynucleotide phosphorylase binds to the 3' end of RNAs and the 3' extension provided by the poly(A) tail allows it to bind to the RNAs whose secondary structure would otherwise block the 3' end. Successive rounds of polyadenylation and degradation of the 3' end by polynucleotide phosphorylase allows the degradosome to overcome these secondary structures. The poly(A) tail can also recruit RNases that cut the RNA in two. These bacterial poly(A) tails are about 30 nucleotides long. In as different groups as animals and trypanosomes, the mitochondria contain both stabilising and destabilising poly(A) tails. Destabilising polyadenylation targets both mRNA and noncoding RNAs. The poly(A) tails are 43 nucleotides long on average. The stabilising ones start at the stop codon, and without them the stop codon (UAA) is not complete as the genome only encodes the U or UA part. Plant mitochondria only have destabilising polyadenylation, and yeast mitochondria have no polyadenylation at all.

While many bacteria and mitochondria have polyadenylate polymerases, they also have another type of polyadenylation, performed by polynucleotide phosphorylase itself. This enzyme is found in bacteria, mitochondria, plastids and as a constituent of the archeal exosome (in those archaea that have an exosome). It can synthesise a 3' extension where the vast majority of the bases are adenines. Like in bacteria, polyadenylation by polynucleotide phosphorylase promotes degradation of the RNA in plastids and likely also archaea.

Evolution

Although polyadenylation is seen in almost all organisms, it is not universal. However, the wide distribution of this modification and the fact that it is present in organisms from all three domains of life implies that the last universal common ancestor of all living organisms probably had some form of polyadenylation system. A few organisms do not polyadenylate mRNA, which implies that they have lost their polyadenylation machineries during evolution. Although no examples of eukaryotes that lack polyadenylation are known, mRNAs from the bacterium *Mycoplasma gallisepticum* and the salt-tolerant archaean *Haloferax volcanii* lack this modification.

The most ancient polyadenylating enzyme is polynucleotide phosphorylase. This enzyme is part of both the bacterial degradosome and archaeal exosome, two closely related complexes that recycle RNA into nucleotides. This enzyme degrades RNA by attacking the bond between the 3'-most nucleotides with a phosphate, breaking off a diphosphate nucleotide.

This reaction is reversible, and so the enzyme can also extend RNA with more nucleotides. The heteropolymeric tail added by polynucleotide phosphorylase is very rich in adenine. The choice of adenine is most likely the result of higher ADP concentrations than other nucleotides as a result of using ATP as an energy currency, making it more likely to be incorporated in this tail in early lifeforms.

It has been suggested that the involvement of adenine-rich tails in RNA degradation prompted the later evolution of polyadenylate polymerases (the enzymes that produce poly(A) tails with no other nucleotides in them). Polyadenylate polymerases are not as ancient. They have separately evolved in both bacteria and eukaryotes from CCA-adding enzyme, which is the enzyme that completes the 3' ends of tRNAs. Its catalytic domain is homologous to that of other polymerases. Likely, the horizontal transfer of bacterial CCA-adding enzyme to eukaryotes allowed the archaeal-like CCA-adding enzyme to switch function to a poly(A) polymerase. Some lineages, like archaea and cyanobacteria, never evolved a polyadenylate polymerase.

History

Polyadenylation was first identified in 1960 as an enzymatic activity in extracts made from cell nuclei that could polymerise ATP, but not ADP, into polyadenine. Although identified in many types of cells, this activity had no known function until 1971, when poly(A) sequences were found in mRNAs. Initially, the only function of these sequences was thought to be protection of the 3' end of the RNA from nucleases, but later the specific roles of polyadenylation in nuclear export and translation were identified. The polymerases responsible for polyadenylation were first purified and characterized in the 1960s and 1970s, but the large number of accessory proteins that control this process were only discovered in the early 1990s.

Polyadenylation is the covalent linkage of a polyadenylyl moiety to a messenger RNA molecule. In eukaryotic organisms, most messenger RNA (mRNA) molecules are polyadenylated at the 3' end.

The poly(A) tail and the protein bound to it aid in protecting mRNA from degradation by exonucleases. Polyadenylation is also important for transcription termination, export of the mRNA from the nucleus, and translation. mRNA can also be polyadenylated in prokaryotic organisms, where poly(A) tails act to facilitate, rather than impede, exonucleolytic degradation.

Polyadenylation occurs during and immediately after transcription of DNA into RNA. After transcription has been terminated, the mRNA chain is cleaved through the action of an endonuclease complex associated with RNA polymerase. After the mRNA has been cleaved, around 250 adenosine residues are added to the free 3' end at the cleavage site. This reaction is catalyzed by polyadenylate polymerase. Just as in alternative splicing, there can be more than one polyadenylation variant of a mRNA.

Transport

Another difference between eukaryotes and prokaryotes is mRNA transport. Because eukaryotic transcription and translation is compartmentally separated, eukaryotic mRNAs must be exported from the nucleus to the cytoplasm. Mature mRNAs are recognized by their processed modifications and then exported through the nuclear pore.

In neurons mRNA must be transported from the soma to the dendrites where local translation occurs in response to external stimuli. Many messages are marked with so-called "zip codes" which targets their transport to a specific location.

Translation

Because prokaryotic mRNA does not need to be processed or transported, translation by the ribosome can begin immediately after the end of transcription. Therefore, it can be said that prokaryotic translation is *coupled* to transcription and occurs *co-transcriptionally*.

Eukaryotic mRNA that has been processed and transported to the cytoplasm (i.e. mature mRNA) can then be translated by the ribosome. Translation may occur at ribosomes free-floating in the cytoplasm, or directed to the endoplasmic reticulum by the signal recognition particle. Therefore, unlike in prokaryotes, eukaryotic translation *is not* directly coupled to transcription.

Structure

5' cap

The 5' cap is a specially altered nucleotide on the 5' end of precursor messenger RNA and some other primary RNA transcripts as found in eukaryotes. The process of 5' capping is vital to creating mature messenger RNA, which is then able to undergo translation. Capping ensures the messenger RNA's stability while it undergoes translation in the process of protein synthesis, and is a highly regulated process that occurs in the cell nucleus.

5' Cap Structure

The 5' cap is found on the 5' end of an mRNA molecule and consists of a guanine nucleotide connected to the mRNA via an unusual 5' to 5' triphosphate linkage. This guanosine is methylated on the 7 position directly after capping *in vivo* by a methyl transferase. It is referred to as a 7-methylguanosine cap, abbreviated m^7G. Further modifications include the possible methylation of the 2' hydroxy-groups of the first 2 ribose sugars of the 5' end of the mRNA. The methylation of both 2' hydroxy-groups is shown on the diagram. The 5' cap looks like the 3' end of an RNA molecule (the 5' carbon of the cap ribose is bonded, and the 3' unbonded). This provides significant resistance to 5' exonucleases.

Capping Process

The starting point is the unaltered 5' end of an RNA molecule. This features a final nucleotide followed by three phosphate groups attached to the 5' carbon.

1. One of the terminal phosphate groups is removed (by RNA terminal phosphatase), leaving two terminal phosphates.
2. GTP is added to the terminal phosphates (by a guanylyl transferase), losing two phosphate groups (from the GTP) in the process. This results in the 5' to 5' triphosphate linkage.
3. The 7-nitrogen of guanine is methylated (by a methyl transferase).
4. Other methyltransferases are optionally used to carry out methylation of 5' proximal nucleotides.

5' Capping Targeting

The Capping Enzyme Complex (CEC) required for capping is found bound to the RNA polymerase II before transcription starts. As soon as the 5' end of the new transcript emerges the enzymes transfer to it and begin the capping process (this is a similar kind of mechanism to ensure capping as for polyadenylation).

The enzymes for capping can only bind to RNA polymerase II ensuring specificity to only these transcripts, which are almost entirely mRNA.

5' Cap Function

The 5' cap has 4 main functions:

1. Regulation of nuclear export.
2. Prevention of degradation by exonucleases.
3. Promotion of translation.
4. Promotion of 5' proximal intron excision.

Nuclear export of RNA is regulated by the Cap binding complex (CBC), which binds exclusively to capped RNA. The CBC is then recognized by the nuclear pore complex and exported. Once in the cytoplasm after the pioneer round of translation, the CBC is replaced by the translation factors eIF-4E and eIF-4G. This complex is then recognized by other translation initiation machinery including the ribosome.

Cap prevents 5' degradation in two ways. First, degradation of the mRNA by 5' exonucleases is prevented (as mentioned above) by functionally looking like a 3' end. Second, the CBC complex and the eIF-4E/eIF-4G block the access of decapping enzymes to the cap. This increases the half-life of the mRNA, essential in eukaryotes as the export process takes significant time.

Decapping of an mRNA is catalyzed by the decapping complex made up of at least Dcp1 and Dcp2, which must compete with eIF-4E to bind the cap. Thus the 5' cap is a marker of an actively translating mRNA and is used by cells to regulate mRNA half-lives in response to new stimuli. Undesirable mRNAs are sent to P-bodies for temporary storage or decapping, the details of which are still being resolved. The mechanism of 5' proximal intron excision promotion is not well understood, but the 5' cap appears to loop around and interact with the spliceosome in the splicing process, promoting intron excision.

Messenger RNA Decapping

The process of messenger RNA decapping consists of hydrolysis of the 5' cap structure on the RNA exposing a 5' monophosphate. This 5' monophosphate is a substrate for the exonuclease Xrn1 and the message is quickly destroyed. There are many situations which may lead to the removal of the cap, some of which are discussed below.

Translation and Decay

Inside cells there is a balance between the processes of translation and mRNA decay. Messages which are being actively translated are bound by polysomes and the initiation factors eIF-4E and eIF-4G. This blocks access to the cap by the decapping enzyme DCP2 and protects the message. In nutrient-starvation conditions or viral infection, translation may be compromised and decapping is stimulated. This balance is reflected in the size and abundance of the cytoplasmic structures known as P-bodies

Specific Decay Pathways

A number of specific decay pathways exist that recognize aberrant messages and promote their decapping. Nonsense mediated decay recognizes premature stop codons and promotes decapping as well as decay by the exosome. Certain classes of miRNA have also been shown to stimulate decapping.

Coding Regions

The coding region of a gene is that portion of a gene's DNA or RNA, composed of exons, that codes for protein. The region is bounded nearer the 5' end by a start codon and nearer the 3' end with a stop codon. The coding region in mRNA is bounded by the five prime untranslated region and the three prime untranslated region, which are also parts of the exons.

The coding region of an organism is sum total of the organism's genome that is composed of gene coding regions. Coding regions are composed of codons, which are decoded and translated (in eukaryotes usually into one and in prokaryotes usually into several) proteins by the ribosome. Coding regions begin with the start codon and end with a stop codon. Generally, the start codon is an AUG triplet and the stop codon is UAA, UAG, or UGA. The coding regions tend to be stabilised by internal base pairs, this impedes degradation. In addition to being protein-coding, portions of coding regions may serve as regulatory sequences in the pre-mRNA as exonic splicing enhancers or exonic splicing silencers.

Open Reading Frame

In molecular genetics, an open reading frame (ORF) is a DNA sequence that does not contain a stop codon in a given reading frame.

Significance

One common use of open reading frames are as one piece of evidence to assist in gene prediction. Long ORFs are often used, along with other evidence, to initially identify candidate protein coding regions in a DNA sequence. The presence of an ORF does not necessarily mean that the region is ever translated. For example in a randomly generated DNA sequence with an equal percentage of each nucleotide, a stop-codon would be expected once every 21 codons. A simple gene prediction algorithm for prokaryotes might look for a start codon followed by an open reading frame that is long enough to encode a typical protein, where the codon usage of that region matches the frequency characteristic for the given organism's coding regions. By itself even a long open reading frame is not conclusive evidence for the presence of a gene.

Example

If a portion of a genome has been sequenced (e.g. 5'-ATCTAAAATGGGTGCC-3'), ORFs can be located by examining each of the three possible reading frames on each strand. In this sequence two out of three possible reading frames are entirely open, meaning that they do not contain a stop codon:

1. ...A TCT AAA ATG GGT GCC...
2. ...AT CTA AAA TGG GTG CC...
3. ...ATC TAA AAT GGG TGC C...

Possible stop codons in DNA are "TGA", "TAA" and "TAG". Thus, the last reading frame in this example contains a stop codon (*TAA*), unlike the first two.

Untranslated Regions

Untranslated regions (UTRs) are sections of the mRNA before the start codon and after the stop codon that are not translated, termed the five prime untranslated region (5' UTR) and three prime untranslated region (3' UTR), respectively. These regions are transcribed with the coding region and thus are exonic as they are present in the mature mRNA. Several roles in gene expression have been attributed to the untranslated regions, including mRNA stability, mRNA localization, and translational efficiency. The ability of a UTR to perform these functions depends on the sequence of the UTR and can differ between mRNAs.

The stability of mRNAs may be controlled by the 5' UTR and/or 3' UTR due to varying affinity for RNA degrading enzymes called ribonucleases and for ancillary proteins that can promote or inhibit RNA degradation.

Translational efficiency, including sometimes the complete inhibition of translation, can be controlled by UTRs. Proteins that bind to either the 3' or 5' UTR may affect translation by influencing the ribosome's ability to bind to the mRNA. MicroRNAs bound to the 3' UTR also may affect translational efficiency or mRNA stability.

Cytoplasmic localization of mRNA is thought to be a function of the 3' UTR. Proteins that are needed in a particular region of the cell can actually be translated there; in such a case, the 3' UTR may contain sequences that allow the transcript to be localized to this region for translation.

Some of the elements contained in untranslated regions form a characteristic secondary structure when transcribed into RNA. These structural mRNA elements are involved in regulating the mRNA. Some, such as the SECIS element, are targets for proteins to bind. One class of mRNA element, the riboswitches, directly bind small molecules, changing their fold to modify levels of transcription or translation. In these cases, the mRNA regulates itself.

5' UTR

The five prime untranslated region (5' UTR), also known as the leader sequence, is a particular section of messenger RNA (mRNA) and the DNA that codes for it. At times the leader sequence can contain elements for controlling gene expression by way of attenuation. The leader sequence will be translated and the availability of the encoded amino acids will affect the transcription of the rest of the mRNA. It starts at the +1 position (where transcription begins) and ends one nucleotide before the start codon (usually AUG) of the coding region. It usually contains a ribosome binding site (RBS), in bacteria also known as the Shine Dalgarno sequence (AGGAGGU). The 5' UTR may be a hundred or more nucleotides long, and the 3' UTR may be even longer (up to several kilobases in length).

An mRNA molecule codes for a protein through translation. The mRNA also contains regions that are not translated: in eukaryotes this includes the 5' untranslated region, 3' untranslated region, 5' cap and poly-A tail.

In prokaryotic mRNA the 5' UTR is normally short. Some viruses and cellular genes have unusual long structured 5' UTRs which may have roles in gene expression.

Several regulatory sequences may be found in the 5' UTR:

- Binding sites for proteins, that may affect the mRNA's stability or translation, for example iron responsive elements, which occur in the 5' UTRs (and 3' UTRs) of a small number of eukaryotic mRNAs that regulate gene expression in response to iron.
- Regulatory elements that do not depend on proteins, such as riboswitches.
- Sequences that promote the initiation of translation.

3' UTR

The three prime untranslated region (3' UTR) is a particular section of messenger RNA (mRNA). It follows the coding region.

An mRNA molecule codes for a protein through translation. The mRNA also contains regions that are not translated. In eukaryotes these regions are the cap, 5' untranslated region, 3' untranslated region, and polyA tail.

In prokaryotes mRNA structures have some differences as do histone mRNAs. However, both have 3' UTRs.

Several regulatory sequences are found in the 3' UTR:

- A polyadenylation signal, usually *AAUAAA*, or a slight variant. This marks the site of cleavage of the transcript approximately 30 base pairs past the signal, followed by the addition of several hundred adenine residues (poly-A tail).
- Binding sites for proteins, that may affect the mRNA's stability or location in the cell, like SECIS elements (which direct the ribosome to translate the codon UGA as selenocysteines rather than as a stop codon), or AU-rich elements (AREs), stretches consisting of mainly adenine and uracil nucleotides (which can either stabilize or destabilize the mRNA depending on the protein bound to it).
- Binding sites for miRNAs, a type of RNAi.

Poly(A) Tail

The 3' poly(A) tail is a long sequence of adenine nucleotides (often several hundred) added to the 3' end of the pre-mRNA. This tail promotes export from the nucleus and translation, and protects the mRNA from degradation.

Monocistronic versus Polycistronic mRNA

An mRNA molecule is said to be monocistronic when it contains the genetic information to translate only a single protein. This is the case for most of the eukaryotic mRNAs.

On the other hand, polycistronic mRNA carries the information of several genes, which are translated into several proteins. These proteins usually have a related function and are grouped and regulated together in an operon. Most of the mRNA found in bacteria and archea are polycistronic. Dicistronic or bicistronic is the term used to describe an mRNA that encodes only two proteins.

mRNA Circularization

In eukaryotes it is thought that mRNA molecules form circular structures due to an interaction between the cap binding complex and poly(A)-binding protein. Circularization is thought to promote recycling of ribosomes on the same message leading to efficient translation.

Degradation

Different mRNAs within the same cell have distinct lifetimes (stabilities). In bacterial cells, individual mRNAs can survive from seconds to more than an hour; in mammalian cells, mRNA lifetimes range from several minutes to days. The greater the stability of an mRNA, the more protein may be produced from that mRNA. The limited lifetime of mRNA enables a cell to alter protein synthesis rapidly in response to its changing needs. There are many mechanisms that lead to the destruction of a mRNA, some of which are described below.

Prokaryotic mRNA Degradation

In prokaryotes the lifetime of mRNA is generally much shorter than in eukaryotes. Prokaryotes degrade messages by using a combination of ribonucleases, including endonucleases, 3' exonucleases, and 5' exonucleases. In some instances, small RNA molecules (sRNA) tens to hundreds of nucleotides long can stimulate the degradation of specific mRNAs by base pairing with complementary sequences and facilitating ribonuclease cleavage. It was recently shown that bacteria also have a sort of 5' cap consisting of a triphosphate on the 5' end. Removal of two of the phosphates leaves a 5' monophosphate, causing the message to be destroyed by the endonuclease RNase E.

Eukaryotic mRNA Turnover

Inside eukaryotic cells there is a balance between the processes of translation and mRNA decay. Messages that are being actively translated are bound by ribosomes, the eukaryotic initiation factors eIF-4E and eIF-4G, and poly(A)-binding protein. eIF-4E and eIF-4G block the decapping enzyme (DCP2), and poly(A)-binding protein blocks the exosome complex, protecting the ends of the message. The balance between translation and decay is reflected in the size and abundance of cytoplasmic structures known as P-bodies The poly(A) tail of the mRNA is shortened by specialized exonucleases that are targeted to specific messenger RNAs by a combination of cis-regulatory sequences on the RNA and trans-acting RNA-binding proteins. Poly(A) tail removal is thought to disrupt the circular structure of the message and destabilize the cap binding complex. The message is then subject to degradation by either the exosome complex or the decapping complex. In this way, translationally inactive messages can be destroyed quickly, while active messages remain intact.

AU-rich Element Decay

The presence of AU-rich elements in some mammalian mRNAs tends to destabilize those transcripts through the action of cellular proteins that bind these sequences and stimulate poly(A) tail removal. Loss of the poly(A) tail is thought to promote mRNA degradation by facilitating attack by both the exosome complex and the decapping complex. Rapid mRNA degradation via AU-rich elements is a critical mechanism for preventing the overproduction of potent cytokines such as tumour necrosis factor (TNF) and granulocyte-macrophage colony stimulating factor (GM-CSF). AU-rich elements also regulate the biosynthesis of proto-oncogenic transcription factors like c-Jun and c-Fos.

Nonsense Mediated Decay

Nonsense-mediated decay (NMD) is a cellular mechanism of mRNA surveillance that functions to detect nonsense mutations and prevent the expression of truncated or erroneous proteins. Following transcription, precursor mRNA undergoes an assemblage of ribonucleoprotein (RNP) components followed by regulatory pre-mRNA processing. Large average intron size in eukaryotic cells greatly increases the probability that aberrant mRNA splicing will result in the presence of a nonsense (stop) codon (UAA, UAG, UGA) somewhere within the open reading frame. NMD is triggered by exon junction complexes (EJCs) (components of the assembled RNP) that are deposited during pre-mRNA processing. While EJCs located in an open reading frame upstream of exon-exon junctions may facilitate ribosomal recruitment prior to displacement by a "pioneer" round of translation, EJCs located downstream of a nonsense codon are not displaced because the ribosome is released from the transcript before reaching it. These remaining EJCs function as tags for recruitment of UPF1 following the mRNA's transport out of the nucleus and into the cytosol where the RNA is degraded, for example by the exosome complex. NMD is not only a mechanism for degrading aberrant mRNA, however, as there are numerous examples of normal transcripts whose expression is regulated by this process including the plasticity protein Arc/Arg3.1.

Eukaryotic messages are subject to surveillance by nonsense mediated decay (NMD), which checks for the presence of premature stop codons (nonsense codons) in the message. These can arise via incomplete splicing, V(D)J recombination in the adaptive immune system, mutations in DNA, transcription errors, leaky scanning by the ribosome causing a frame shift, and other causes. Detection of a premature stop codon triggers mRNA degradation by 5' decapping, 3' poly(A) tail removal, or endonucleolytic cleavage.

Small Interfering RNA (siRNA)

Small interfering RNA (siRNA), sometimes known as short interfering RNA or silencing RNA, is a class of double-stranded RNA molecules, 20-25 nucleotides in length, that play a variety of roles in biology. The most notable role of siRNA is its involvement in the RNA interference (RNAi) pathway, where it interferes with the expression of a specific gene. In addition to its role in the RNAi pathway, siRNA also acts in RNAi-related pathways, e.g., as an antiviral mechanism or in shaping the chromatin structure of a genome; the complexity of these pathways is only now being elucidated.

siRNAs were first discovered by David Baulcombe's group at the Sainsbury Laboratory in Norwich, England, as part of post-transcriptional gene silencing (PTGS) in plants. The group published their findings in *Science* in a paper titled "A species of small antisense RNA in posttranscriptional gene silencing in plants". Shortly thereafter, in 2001, synthetic siRNAs were shown to be able to induce RNAi in mammalian cells by Thomas Tuschl, and colleagues in a paper published in *Nature*. This discovery led to a surge in interest in harnessing RNAi for biomedical research and drug development.

RNAi Induction Using siRNAs or their Biosynthetic Precursors

Transfection of an exogenous siRNA can be problematic because the gene knockdown effect is only transient, in particular, in rapidly dividing cells. One way of overcoming this challenge is to modify the siRNA in such a way as to allow it to be expressed by an appropriate vector, e.g., a plasmid.

This is done by the introduction of a loop between the two strands, thus producing a single transcript, which can be processed into a functional siRNA. Such transcription cassettes typically use an RNA polymerase III promoter (e.g., U6 or H1), which usually directs the transcription of small nuclear RNAs (snRNAs) (U6 is involved in gene splicing; H1 is the RNase component of human RNase P). It is assumed (although not known for certain) that the resulting siRNA transcript is then processed by Dicer.

RNA Activation

It has recently been found that dsRNA can also activate gene expression, a mechanism that has been termed "small RNA-induced gene activation" or RNAa. It has been shown that dsRNAs targeting gene promoters induce potent transcriptional activation of associated genes. RNAa was demonstrated in human cells using synthetic dsRNAs, termed "small activating RNAs" (saRNAs). It is currently not known whether RNAa is conserved in other organisms.

Challenges: Avoiding Nonspecific Effects

Because RNAi intersects with a number of other pathways, it is not surprising that on occasion nonspecific effects are triggered by the experimental introduction of an siRNA. When a mammalian cell encounters a double-stranded RNA such as an siRNA, it may mistake it as a viral by-product and mount an immune response. Furthermore, because structurally related microRNAs modulate gene expression largely via incomplete complementarity base pair interactions with a target mRNA, the introduction of an siRNA may cause unintended off-targeting.

Innate Immunity

Introduction of too much siRNA can result in nonspecific events due to activation of innate immune responses. Most evidence to date suggests that this is probably due to activation of the dsRNA sensor PKR, although retinoic acid-inducible gene I (RIG-I) may also be involved. The induction of cytokines via toll-like receptor 7 (TLR7) has also been described. One promising method of reducing the nonspecific effects is to convert the siRNA into a microRNA. MicroRNAs occur naturally, and by harnessing this endogenous pathway it should be possible to achieve similar gene knockdown at comparatively low concentrations of resulting siRNAs. This should minimize nonspecific effects.

Off-targeting

Off-targeting is another challenge to the use of siRNAs as a gene knockdown tool. Here, genes with incomplete complementarity are inadvertently downregulated by the siRNA (in effect, the siRNA acts as a miRNA), leading to problems in data interpretation and potential toxicity.

This, however, can be partly addressed by designing appropriate control experiments, and siRNA design algorithms are currently being developed to produce siRNAs free from off-targeting. Genome-wide expression analysis, e.g., by microarray technology, can then be used to verify this and further refine the algorithms. A 2006 paper from the laboratory of Dr. Khvorova implicates 6- or 7-basepair-long stretches from position 2 onward in the siRNA matching with 3'UTR regions in off-targeted genes.

Possible Therapeutic Applications and Challenges

Given the ability to knock down, in essence, any gene of interest, RNAi via siRNAs has generated a great deal of interest in both basic and applied biology. There are an increasing number of large-scale RNAi screens that are designed to identify the important genes in various biological pathways. Because disease processes also depend on the activity of multiple genes, it is expected that in some situations turning off the activity of a gene with an siRNA could produce a therapeutic benefit.

However, applying RNAi via siRNAs to living animals, especially humans, poses many challenges. Under experiments, siRNAs show different effectiveness in different cell types in a manner as yet poorly understood: Some cells respond well to siRNAs and show a robust knockdown, whereas others show no such knockdown (even despite efficient transfection).

Phase I results of the first two therapeutic RNAi trials (indicated for age-related macular degeneration, aka AMD) reported at the end of 2005 that siRNAs are well tolerated and have suitable pharmacokinetic properties.

Proof of concept trials have indicated that Ebola-targeted siRNAs may be effective as post-exposure prophylaxis in humans, with 100% of non-human primates surviviing a lethal dose of Zaire Ebolavirus, the most lethal strain.

In metazoans, small interfering RNAs (siRNAs) processed by Dicer are incorporated into a complex known as the RNA-induced silencing complex or RISC. This complex contains an endonuclease that cleaves perfectly complementary messages to which the siRNA binds. The resulting mRNA fragments are then destroyed by exonucleases. siRNA is commonly used in laboratories to block the function of genes in cell culture. It is thought to be part of the innate immune system as a defense against double-stranded RNA viruses.

MicroRNA (miRNA)

MicroRNAs (miRNAs) are small RNAs that typically are partially complementary to sequences in metazoan messenger RNAs. Binding of a miRNA to a message can repress translation of that message and accelerate poly(A) tail removal, thereby hastening mRNA degradation. The mechanism of action of miRNAs is the subject of active research. MicroRNAs (miRNAs) are short ribonucleic acid (RNA) molecules, on average only 22 nucleotides long and are found in all eukaryotic cells, except fungi and marine plants. miRNAs are post-transcriptional regulators that bind to complementary sequences on target messenger RNA transcripts (mRNAs), usually resulting in translational repression and gene silencing. The human genome may encode over 1000 miRNAs, which may target about 60% of mammalian genes and are abundant in many human cell types.

miRNAs show very different characteristics between plants and metazoans. In plants the miRNA complementarity to its mRNA target is nearly perfect, with no or few mismatched bases. In metazoans on the other hand miRNA complementarity is far from perfect and one miRNA can target many different sites on the same mRNA or on many different mRNAs. Another difference is the location of target sites on mRNAs. In metazoans the miRNA target sites are in the three prime untranslated regions (3'UTR) of the mRNA. In plants targets can be located in the 3' UTR but are more often in the coding region itself. MiRNAs are well conserved in eukaryotic organism and are thought to be a vital and evolutionarily ancient component of genetic regulation.

The first miRNAs were characterized in the early 1990s, but miRNAs were not recognized as a distinct class of biologic regulators with conserved functions until the early 2000s. Since then, miRNA research has revealed multiple roles in negative regulation (transcript degradation and sequestering, translational suppression) and possible involvement in positive regulation (transcriptional and translational activation). By affecting gene regulation, miRNAs are likely to be involved in most biologic processes. Different sets of expressed miRNAs are found in different cell types and tissues.

Aberrant expression of miRNAs has been implicated in numerous disease states, and miRNA-based therapies are under investigation.

History

MicroRNAs were discovered in 1993 by Victor Ambros, Rosalind Lee and Rhonda Feinbaum during a study of the gene *lin-14* in *C. elegans* development. They found that LIN-14 protein abundance was regulated by a short RNA product encoded by the *lin-4* gene. A 61 nucleotide precursor from *lin-4* gene matured to a 22 nucleotide RNA containing sequences partially complementary to multiple sequences in the 3' UTR of the *lin-14* mRNA. This complementarity was sufficient and necessary to inhibit the translation of *lin-14* mRNA into LIN-14 protein. Retrospectively, the *lin-4* small RNA was the first microRNA to be identified, though at the time, it was thought to be a nematode idiosyncrasy. Only in 2000 was a second RNA characterized: let-7, which repressed *lin-41*, *lin-14*, *lin-28*, *lin-42*, and *daf-12* expression during developmental stage transitions in *C. elegans*. let-7 was soon found to be conserved in many species, indicating the existence of a wider phenomenon.

Nomenclature

Under a standard nomenclature system, names are assigned to experimentally confirmed miRNAs before publication of their discovery. The prefix "mir" is followed by a dash and a number, the latter often indicating order of naming. For example, mir-123 was named and likely discovered prior to mir-456. The uncapitalized "mir-" refers to the pre-miRNA, while a capitalized "miR-" refers to the mature form. miRNAs with nearly identical sequences bar one or two nucleotides are annotated with an additional lower case letter. For example, miR-123a would be closely related to miR-123b. Pre-miRNAs that lead to 100% identical mature miRNAs but that are located at different places in the genome are indicated with an additional dash-number suffix.

For example, the pre-miRNAs hsa-mir-194-1 and hsa-mir-194-2 lead to an identical mature miRNA (hsa-miR-194) but are located in different regions of the genome. Species of origin is designated with a three-letter prefix, e.g., hsa-miR-123 would be from human (*Homo sapiens*) and oar-miR-123 would be a sheep (*Ovis aries*) miRNA. Other common prefixes include 'v' for viral (miRNA encoded by a viral genome) and 'd' for *Drosophila* miRNA (a fruit fly commonly studied in genetic research). When two mature microRNAs originate from opposite arms of the same pre-miRNA, they are denoted with a -3p or -5p suffix. (In the past, this distinction was also made with 's' (sense) and 'as' (antisense)). When relative expression levels are known, an asterisk following the name indicates an miRNA expressed at low levels relative to the miRNA in the opposite arm of a hairpin. For example, miR-123 and miR-123* would share a pre-miRNA hairpin, but more miR-123 would be found in the cell.

Biogenesis

Most microRNA genes are found in intergenic regions or in anti-sense orientation to genes and contain their own miRNA gene promoter and regulatory units. As much as 40% of miRNA genes may lie in the introns of protein and non-protein coding genes or even in exons.

These are usually, though not exclusively, found in a sense orientation. and thus usually are regulated together with their host genes. Other miRNA genes showing a common promoter include the 42-48% of all miRNAs originating from polycistronic units containing 2-7 discrete loops from which mature miRNAs are processed, although this does not necessarily mean the mature miRNAs of a family will be homologous in structure and function. The promoters mentioned have been shown to have some similarities in their motifs to promoters of other genes transcribed by RNA polymerase II such as protein coding genes. The DNA template is not the final word on mature miRNA production: 6% of human miRNAs show RNA editing, the site-specific modification of RNA sequences to yield products different from those encoded by their DNA. This increases the diversity and scope of miRNA action beyond that implicated from the genome alone.

Transcription

miRNA genes are usually transcribed by RNA polymerase II (Pol II). The polymerase often binds to a promoter found near the DNA sequence encoding what will become the hairpin loop of the pre-miRNA. The resulting transcript is capped with a specially-modified nucleotide at the 5' end, polyadenylated with multiple adenosines (a poly(A) tail), and spliced. The product, called a primary miRNA (pri-miRNA), may be hundreds or thousands of nucleotides in length and contain one or more miRNA stem loops. When a stem loop precursor is found in the 3' UTR, a transcript may serve as a pri-miRNA and a mRNA. RNA polymerase III (Pol III) transcribes some miRNAs, especially those with upstream Alu sequences, transfer RNAs (tRNAs), and mammalian wide interspersed repeat (MWIR) promoter units.

Nuclear Processing

A single pri-miRNA may contain from one to six miRNA precursors. These hairpin loop structures are composed of about 70 nucleotides each. Each hairpin is flanked by sequences necessary for efficient processing. The double-stranded RNA structure of the hairpins in a pri-miRNA is recognized by a nuclear protein known as DiGeorge Syndrome Critical Region 8 (DGCR8 or

"Pasha" in invertebrates), named for its association with DiGeorge Syndrome. DGCR8 associates with the enzyme Drosha, a protein that cuts RNA, to form the "Microprocessor" complex. In this complex, DGCR8 orients the catalytic RNase III domain of Drosha to liberate hairpins from pri-miRNAs by cleaving RNA about eleven nucleotides from the hairpin base (two helical RNA turns into the stem). The resulting hairpin, known as a pre-miRNA (precursor-miRNA), has a two-nucleotide overhang at its 3' end; it has 3' hydroxyl and 5' phosphate groups.

Pre-miRNAs that are spliced directly out of introns, bypassing the Microprocessor complex, are known as "mirtrons." Originally thought to exist only in *Drosophila* and *C. elegans*, mirtrons have now been found in mammals.

Perhaps as many as 16% of pri-miRNAs may be altered through nuclear RNA editing. Most commonly, enzymes known as adenosine deaminases acting on RNA (ADARs) catalyze adenosine to inosine (A to I) transitions. RNA editing can halt nuclear processing (for example, of pri-miR-142, leading to degradation by the ribonuclease Tudor-SN) and alter downstream processes including cytoplasmic miRNA processing and target specificity (e.g., by changing the seed region of miR-376 in the central nervous system).

Nuclear Export

pre-miRNA hairpins are exported from the nucleus in a process involving the nucleocytoplasmic shuttle Exportin-5. This protein, a member of the *karyopherin* family, recognizes a two-nucleotide overhang left by the RNase III enzyme Drosha at the 3' end of the pre-miRNA hairpin. Exportin-5-mediated transport to the cytoplasm is energy-dependent, using GTP bound to the Ran protein.

Cytoplasmic Processing

In the cytoplasm, the pre-miRNA hairpin is cleaved by the RNase III enzyme Dicer. This endoribonuclease interacts with the 3' end of the hairpin and cuts away the loop joining the 3' and 5' arms, yielding an imperfect miRNA:miRNA* duplex about 22 nucleotides in length. Overall hairpin length and loop size influence the efficiency of Dicer processing, and the imperfect nature of the miRNA:miRNA* pairing also affects cleavage. Although either strand of the duplex may potentially act as a functional miRNA, only one strand is usually incorporated into the RNA-induced silencing complex (RISC) where the miRNA and its mRNA target interact.

Biogenesis In Plants

miRNA biogenesis in plants differs from metazoan biogenesis mainly in the steps of nuclear processing and export. Instead of being cleaved by two different enzymes, once inside and once outside the nucleus, both cleavages of the plant miRNA is performed by a Dicer homolog, called Dicer-like1 (DL1). DL1 is only expressed in the nucleus of plant cells, which indicates that both reactions take place inside the nucleus. Before plant miRNA:miRNA* duplexes are transported out of the nucleus its 3' overhangs are methylated by a RNA methyltransferaseprotein called Hua-Enhancer1 (HEN1). The duplex is then transported out of the nucleus to the cytoplasm by a protein called Hasty (HST), an Exportin 5 homolog, where they disassemble and the mature miRNA is incorporated into the RISC.

The RNA-induced Silencing Complex

The mature miRNA is part of an active RNA-induced silencing complex (RISC) containing Dicer and many associated proteins. RISC is also known as a microRNA ribonucleoprotein complex (miRNP); RISC with incorporated miRNA is sometimes referred to as "miRISC."

Dicer processing of the pre-miRNA is thought to be coupled with unwinding of the duplex. Generally, only one strand is incorporated into the miRISC, selected on the basis of its thermodynamic instability and weaker base-pairing relative to the other strand. The position of the stem-loop may also influence strand choice. The other strand, called the passenger strand due to its lower levels in the steady state, is denoted with an asterisk (*) and is normally degraded. In some cases, both strands of the duplex are viable and become functional miRNA that target different mRNA populations.

Members of the argonaute (Ago) protein family are central to RISC function. Argonautes are needed for miRNA-induced silencing and contain two conserved RNA binding domains: a PAZ domain that can bind the single stranded 3' end of the mature miRNA and a PIWI domain that structurally resembles ribonuclease-H and functions to interact with the 5' end of the guide strand.

They bind the mature miRNA and orient it for interaction with a target mRNA. Some argonautes, for example human Ago2, cleave target transcripts directly; argonautes may also recruit additional proteins to achieve translational repression. The human genome encodes eight argonaute proteins divided by sequence similarities into two families: AGO (with four members present in all mammalian cells and called E1F2C/hAgo in humans), and PIWI (found in the germ line and hematopoietic stem cells).

Additional RISC components include TRBP [human immunodeficiency virus (HIV) transactivating response RNA (TAR) binding protein], PACT (protein activator of the interferon induced protein kinase (PACT), the SMN complex, fragile X mental retardation protein (FMRP), and Tudor staphylococcal nuclease-domain-containing protein (Tudor-SN).

Mode of Silencing

Gene silencing may occur either via mRNA degradation or preventing mRNA from being translated. It has been demonstrated that if there is complete complementation between the miRNA and target mRNA sequence, Ago2 can cleave the mRNA and lead to direct mRNA degradation. Yet, if there isn't complete complementation the silencing is achieved by preventing translation.

miRNA Turnover

Turnover of mature miRNA is needed for rapid changes in miRNA expression profiles. During miRNA maturation in the cytoplasm, uptake by the Argonaute protein is thought to stabilize the guide strand, while the opposite (* or "passenger") strand is preferentially destroyed. In what has been called a "Use it or lose it" strategy, Argonaute may preferentially retain miRNAs with many targets over miRNAs with few or no targets, leading to degradation of the non-targeting molecules.

Decay of mature miRNAs in animals is mediated by the 5′-to-3′ exoribonuclease XRN2, also known as Rat1p. In plants, SDN (small RNA degrading nuclease) family members degrade miRNAs in the opposite (3'-to-5') direction. Similar enzymes are encoded in animal genomes, but their roles have not yet been described.

Several miRNA modifications affect miRNA stability. As indicated by work in the model organism *Arabidopsis thaliana* (thale cress), mature plant miRNAs appear to be stabilized by the addition of methyl moieties at the 3' end. The 2'-O-conjugated methyl groups block the addition of uracil (U) residues by uridyltransferase enzymes, a modification that may be associated with miRNA degradation. However, uridylation may also protect some miRNAs; the consequences of this modification are incompletely understood. Uridylation of some animal miRNAs has also been reported. Both plant and animal miRNAs may be altered by addition of adenine (A) residues to the 3' end of the miRNA. An extra A added to the end of mammalian miR-122, a liver-enriched miRNA important in Hepatitis C, stabilizes the molecule, and plant miRNAs ending with an adenine residue have slower decay rates.

Cellular Functions

The function of miRNAs appears to be in gene regulation. For that purpose, a miRNA is complementary to a part of one or more messenger RNAs (mRNAs). Animal miRNAs are usually complementary to a site in the 3' UTR whereas plant miRNAs are usually complementary to coding regions of mRNAs. Perfect or near perfect base pairing with the target RNA promotes cleavage of the RNA. This is the primary mode of plant microRNAs. In animals, microRNAs more often only partially base pair and inhibit protein translation of the target mRNA (this exists in plants as well but is less common). MicroRNAs that are partially complementary to a target can also speed up deadenylation, causing mRNAs to be degraded sooner. For partially complementary microRNAs to recognise their targets, nucleotides 2–7 of the miRNA (its 'seed region') still have to be perfectly complementary. miRNAs occasionally also cause histone modification and DNA methylation of promoter sites, which affects the expression of target genes.

Animal microRNAs target in particular developmental genes. In contrast, genes involved in functions common to all cells, such as gene expression, have very few microRNA target sites and seem to be under selection to avoid targeting by microRNAs.

dsRNA can also activate gene expression, a mechanism that has been termed "small RNA-induced gene activation" or RNAa. dsRNAs targeting gene promoters can induce potent transcriptional activation of associated genes. This was demonstrated in human cells using synthetic dsRNAs termed small activating RNAs (saRNAs), but has also been demonstrated for endogenous microRNA.

Evolution

MicroRNAs are significant phylogenetic markers because of their astonishingly low rate of evolution. Their origin may have permitted the development of morphological innovation, and by making gene expression more specific and 'fine-tunable', permitted the genesis of complex organs and perhaps, ultimately, complex life. Indeed, rapid bursts of morphological innovation are generally associated with a high rate of microRNA accumulation.

MicroRNAs originate predominantly by the random formation of hairpins in "non-coding" sections of DNA (i.e. introns or intergene regions), but also by the duplication and modification of existing microRNAs. The rate of evolution (i.e. nucleotide substitution) in recently-originated microRNAs is comparable to that elsewhere in the non-coding DNA, implying evolution by neutral drift; however, older microRNAs have a much lower rate of change (often less than one substitution per hundred million years), suggesting that once a microRNA gains a function it undergoes extreme purifying selection.

At this point, a microRNA is rarely lost from an animal's genome, although microRNAs which are more recently derived (and thus presumably non-functional) are frequently lost. This makes them a valuable phylogenetic marker, and they are being looked upon as a possible solution to such outstanding phylogenetic problems as the relationships of arthropods.

MicroRNAs feature in the genomes of most eukaryotic organisms, from the brown algae to the metazoa. Across all species, in excess of 5000 had been identified by March 2010. Whilst short RNA sequences (50 – hundreds of base pairs) of a broadly comparable function occur in bacteria, bacteria lack true microRNAs.

Experimental Detection and Manipulation of miRNA

MicroRNA expression can be quantified in a two-step polymerase chain reaction process of modified RT-PCR followed by quantitative real-time PCR. Variations of this method achieve absolute or relative quantification. miRNAs can also be hybridized to microarrays, slides or chips with probes to hundreds or thousands of miRNA targets, so that relative levels of miRNAs can be determined in different samples. MicroRNAs can be both discovered and profiled by high-throughput sequencing methods. The activity of an miRNA can be experimentally inhibited using a locked nucleic acid (LNA) oligo, a Morpholino oligo or a 2'-O-methyl RNA oligo. MicroRNA maturation can be inhibited at several points by steric-blocking oligos. The miRNA target site of an mRNA transcript can also be blocked by a steric-blocking oligo. Additionally, a specific miRNA can be silenced by a complementary antagomir. For the "in situ" detection of miRNA, the use of LNA is currently the only efficient method. The locked conformation of LNA results in enhanced hybridization properties and increases sensitivity and selectivity, making it ideal for detection of short miRNA.

miRNA and Disease

Just as miRNA is involved in the normal functioning of eukaryotic cells, so has dysregulation of miRNA been associated with disease. A manually curated, publicly available database miR2Disease documents known relationships between miRNA dysregulation and human disease.

miRNA and Cancer

Several miRNAs have been found to have links with some types of cancer. A study of mice altered to produce excess c-Myc — a protein with mutated forms implicated in several cancers — shows that miRNA has an effect on the development of cancer. Mice that were engineered to produce a surplus of types of miRNA found in lymphoma cells developed the disease within 50 days and died two weeks later. In contrast, mice without the surplus miRNA lived over 100 days. Leukemia can be caused by the insertion of a viral genome next to the 17-92 array of microRNAs leading to increased expression of this microRNA.

Another study found that two types of miRNA inhibit the E2F1 protein, which regulates cell proliferation. miRNA appears to bind to messenger RNA before it can be translated to proteins that switch genes on and off. By measuring activity among 217 genes encoding miRNA, patterns of gene activity that can distinguish types of cancers can be discerned. miRNA signatures may enable classification of cancer. This will allow doctors to determine the original tissue type which spawned a cancer and to be able to target a treatment course based on the original tissue type. miRNA profiling has already been able to determine whether patients with chronic lymphocytic leukemia had slow growing or aggressive forms of the cancer.

Transgenic mice that over-express or lack specific miRNAs have provided insight into the role of small RNAs in various malignancies.

A novel miRNA-profiling based screening assay for the detection of early-stage colorectal cancer has been developed and is currently in clinical trials. Early results showed that blood plasma samples collected from patients with early, resectable (Stage II) colorectal cancer could be distinguished from those of sex-and age-matched healthy volunteers. Sufficient selectivity and specificity could be achieved using small (less than 1 mL) samples of blood. The test has potential to be a cost-effective, non-invasive way to identify at-risk patients who should undergo colonoscopy.

miRNA and Heart Disease

The global role of miRNA function in the heart has been addressed by conditionally inhibiting miRNA maturation in the murine heart, and has revealed that miRNAs play an essential role during its development. miRNA expression profiling studies demonstrate that expression levels of specific miRNAs change in diseased human hearts, pointing to their involvement in cardiomyopathies. Furthermore, studies on specific miRNAs in animal models have identified distinct roles for miRNAs both during heart development and under pathological conditions, including the regulation of key factors important for cardiogenesis, the hypertrophic growth response, and cardiac conductance.

miRNA and the Nervous System

miRNAs appear to regulate the nervous system. Neural miRNAs are involved at various stages of synaptic development, including dendritogenesis (involving miR-132, miR-134 and miR-124), synapse formation and synapse maturation (where miR-134 and miR-138 are thought to be involved). Some studies find altered miRNA expression in schizophrenia.

miRNA and Non-coding RNAs

When the human genome project mapped its first chromosome in 1999, it was predicted the genome would contain over 100,000 protein coding genes. However, only around 20,000 were eventually identified (International Human Genome Sequencing Consortium, 2004). Since then, the advent of bioinformatics approaches combined with genome tiling studies examining the transcriptome, systematic sequencing of full length cDNA libraries, and experimental validation (including the creation of miRNA derived antisense oligonucleotides called antagomirs) have revealed that many transcripts are non protein-coding RNA, including several snoRNAs and miRNAs.

Other Decay Mechanisms

There are other ways in which messages can be degraded, including non-stop decay, silencing by Piwi-interacting RNA (piRNA), and surely other means.

7

Chromosome Mutations

It is natures intention that the exact genetic information from both parents will be seen in the offspring's DNA in the the critical stages of fertilisation. However, it is possible for this genetic information to mutate, which in most cases, can result in fatal or negative consequencies in the outcome of the new ogranism.

Non-Disjunction and Down's Syndrome

One well known example of mutation is non-disjunction. Non-disjunction is when the spindle fibres fail to seperate during meiosis, resulting in gametes with one extra chromosome and other gametes lacking a chromosome.

If this non-disjunction occurs in chromosome 21 of a human egg cell, a condition called Down's syndrome occurs. This is because their cells possess 47 chromosomes as opposed to the normal chromosome compliment in humans of 46.

The fundamental structure of a chromosome is subject to mutation, which will most likely occur during crossing over at meiosis. There are a number of ways in which the chromosome structure can change, as indicated below, which will detrimentally change the genotype and phenotype of the organism.

However, if the chromosome mutation effects an essential part of DNA, it is possible that the mutation will abort the offspring before it has the chance of being born. The following indicates types of chromosome mutation where whole genes are moved:

Deletion of a Gene

As the name implies, genes of a chromosome are permanently lost as they become unattached to the centromere and are lost forever

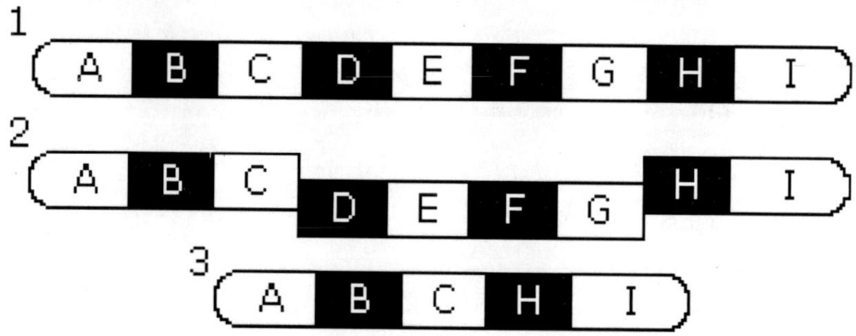

- Normal chromosome before mutation
- Genes not attached to centromere become loose and lost forever
- New chromosome lacks certain genes which may prove fatal depending on how important these genes are

Duplication of Genes

In this mutation, the mutants genes are displayed twice on the same chromosome due to duplication of these genes. This can prove to be an advantageous mutation as no genetic information is lost or altered and new genes are gained

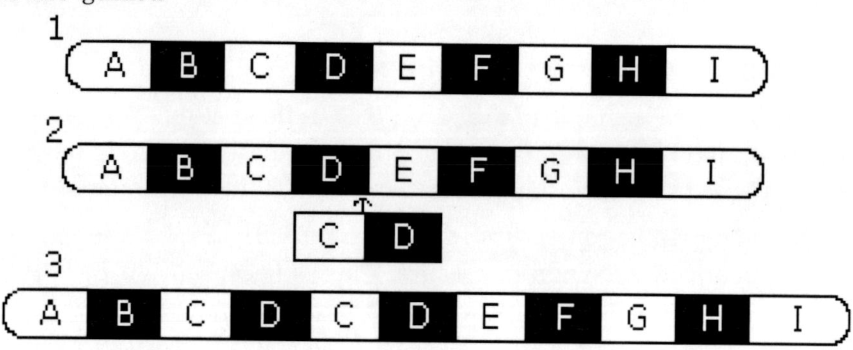

- Normal chromosome before mutation
- Genes from the homologous chromosome are copied and inserted into the genetic sequence
- New chromosome possesses all its initial genes plus a duplicated one, which is usually harmless

Inversion of Genes

This is where the order of a particular order of genes are reversed as seen below

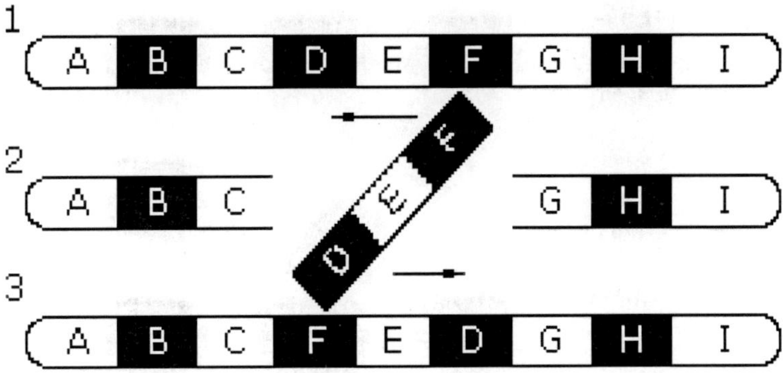

- Normal chromosome un-altered
- The connection between genes break and the sequence of these genes are reversed
- The new sequence may not be viable to produce an organism, depending on which genes are reversed. Advantageous characteristics from this mutation are also possible

Translocation of Genes

This is where information from one of two homologous chromosomes breaks and binds to the other. Usually this sort of mutation is lethal

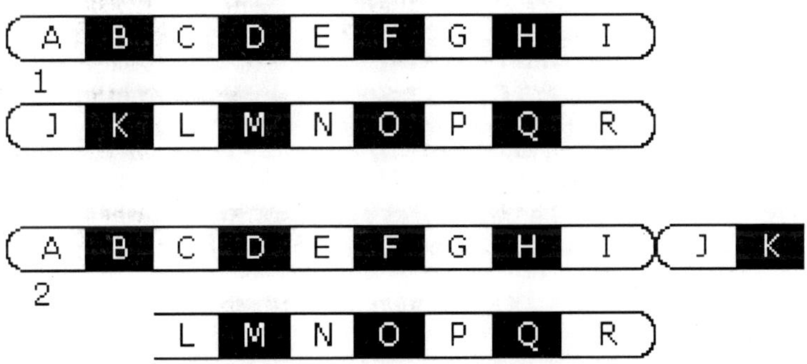

- An un-altered pair of homologous chromosomes
- Translocation of genes has resulted in some genes from one of the chromosomes attaching to the opposing chromosome

Alteration of a DNA Sequence

The previous examples of mutation have investigated changes at the chromosome level. The sequence of nucleotides on a DNA sequence are also susceptible to mutation.

Deletion: Here, certain nucleotides are deleted, which affects the coding of proteins that use this DNA sequence. If for example, a gene coded for alanine, with a genetic sequence of C-G-G, and the cytosine nucleotide was deleted, then the alanine amino acid would not be able to be created, and any other amino acids that are supposed to be coded from this DNA sequence will also be unable to be produced because each successive nucleotide after the deleted nucleotide will be out of place.

Insertion: Similar to the effects of deletion, where a nucleotide is inserted into a genetic sequence and therefore alters the chain thereafter. This alteration of a nucleotide sequence is known as frameshift

Inversion: Where a particular nucleotide sequence is reversed, and is not as serious as the above mutations. This is because the nucleotides that have been reversed in order only affect a small portion of the sequence at large

Substitution: A certain nucleotide is replaced with another, which will affect any amino acid to be synthesised from this sequence due to this change. If the gene is essential, i.e. for the coding of haemoglobin then the effects are serious, and organisms in this instance suffer from a condition called sickle cell anaemia.

All of the genetic mutations looked at through the last 2 pages more or less have a negative impact and are undesired, however, in some cases they can prove advantageous.

Genetic mutations increase genetic diversity and therefore have an important part to play. They are also the reason many people inherit diseases.

Mutation Frequency and Polyploidy

The previous two pages have investigated mutations, and this page continues with more information related to genetic mutations.

Polyploidy

Humans are diploid creatures, meaning for every chromosome in our body, there is another one to match it. Read the following

- Haploid creatures have one of each chromosome
- Diploid creatures have two of each chromosome
- Triploid creatures have three of each chromosome
- Polyploid creatures have three or more of each chromosome
- They can be represented by n where n equals haploid, 2n equals diploid and so on.

It is possible for a species, particularly plant species, to produce offspring that contains more chromosomes than its parent. This can be a result of non-disjunction, where normally a diploid parent would produce diploid offspring, but in the case of non-disjunction in one of the parents, produces a polyploid.

In the case of triploids, although the creation of particular triploids in species is possible, they cannot reproduce themselves because of the inability to pair homologous chromosomes at meiosis, therefore preventing the formation of gametes. Polyploidy is responsible for the creation of thousands of species in today's planet, and will continue to do so. It is also responsible for increasing genetic diversity and producing species showing an increase in size, vigour and an increased resistance to disease.

Mutation Frequency

This page and the previous two have investigated the different ways that mutations arise, and the following elaborates on the ways in which mutations are instigated.Barring all external factors, mutations occur very rarely, and are rarely expressed because many forms of mutation are expressed by a recessive allele. However there are many mutagenic agents that artificially increase the rate of mutations in an organism. The following are some factors that increase genetic mutations in organisms

- Members of species in a particular geographic area or ethnic origin are more susceptible to mutations
- High dosages of X-Rays or ultraviolet light can increase the likeliness of a mutation
- Radioactive substances increase the rate of mutations exponentially

As mentioned previously, genetic mutations are a source of new variation in a species because it physically alters the sequence of nucleotides in a given sequence, therefore altering the genome in a unique way.

Theory of Natural Selection

In the 19th century, a man called Charles Darwin, a biologist from England, set off on the ship HMS Beagle to investigate species of the island. After spending time on the islands, he soon developed a theory that would contradict the creation of man and imply that all species derived from common ancestors through a process called natural selection. Natural selection is considered to be the biggest factor resulting in the diversity of species and their genomes. The principles of Darwin's work and his theory are stated below.

- One of the prime motives for all species is to reproduce and survive, passing on the genetic information of the species from generation to generation. When species do this they tend to produce more offspring than the environment can support.

- The lack of resources to nourish these individuals places pressure on the size of the species population, and the lack of resources means increased competition and as a consequence, some organisms will not survive.
- The organisms who die as a consequence of this competition were not totally random, Darwin found that those organisms more suited to their environment were more likely to survive.
- This resulted in the well known phrase survival of the fittest, where the organisms most suited to their environment had more chance of survival if the species falls upon hard times. (This phrase if often associated with Darwin, though on closer inspection Herbert Spencer puts the phrase in a more accurate historical context.)
- Those organisms who are better suited to their environment exhibit desirable characteristics, which is a consequence of their genome being more suitable to begin with.

This 'weeding out' of less suited organisms and the reward of survival to those better suited led Darwin to deduce that organisms had evolved over time, where the most desirable characteristics of a species are favoured and those organisms who exhibit them survive to pass their genes on. As a consequence of this, a changing environment would mean different characteristics would be favourable in a changing environment.

Darwin believed that organisms had 'evolved' to suit their environments, and occupy an ecological niche where they would be best suited to their environment and therefore have the best chance of survival.

As the above indicates, those alleles of a species that are favoured in the environment will become more frequent in the genomes of the species, due to the organisms higher likeliness of surviving as part of the species at large

Recombinant DNA

Recombinant DNA (rDNA) molecules are DNA sequences that result from the use of laboratory methods (molecular cloning) to bring together genetic material from multiple sources, creating sequences that would not otherwise be found in biological organisms. Recombinant DNA is possible because DNA molecules from all organisms share the same chemical structure; they differ only in the sequence of nucleotides within that identical overall structure.

Consequently, when DNA from a foreign source is linked to host sequences that can drive DNA replication and then introduced into a host organism, the foreign DNA is replicated along with the host DNA. Recombinant DNA molecules are sometimes called chimeric DNA, because they are usually made of material from two different species, like the mythological chimera.

The DNA sequences used in the construction of recombinant DNA molecules can originate from any species. For example, plant DNA may be joined to bacterial DNA, or human DNA may be joined with fungal DNA. In addition, DNA sequences that do not occur anywhere in nature may be created by the chemical synthesis of DNA, and incorporated into recombinant molecules.

Using recombinant DNA technology and synthetic DNA, literally any DNA sequence may be created and introduced into any of a very wide range of living organisms.

Proteins that result from the expression of recombinant DNA within living cells are termed recombinant proteins. When recombinant DNA encoding a protein is introduced into a host organism, the recombinant protein will not necessarily be produced .

Expression of foreign proteins requires the use of specialized expression vectors and often necessitates significant restructuring of the foreign coding sequence. It is important to note that recombinant DNA differs from genetic recombination in that the former results from artificial methods in the test tube, while the latter is a normal biological process that results in the remixing of existing DNA sequences in essentially all organisms.

Creating Recombinant DNA

Construction of recombinant DNA, in which a foreign DNA fragment is inserted into a plasmid vector. In this example, the gene indicated by the white color is inactivated upon insertion of the foreign DNA fragment.

Molecular cloning is the laboratory process used to create recombinant DNA. It is one of two widely-used methods (along with polymerase chain reaction, abbr.

PCR) used to direct the replication of any specific DNA sequence chosen by the experimentalist. The fundamental difference between the two methods is that molecular cloning involves replication of the DNA within a living cell, while PCR replicates DNA in the test tube, free of living cells.

Formation of recombinant DNA requires a cloning vector, a DNA molecule that will replicate within a living cell. Vectors are generally derived from plasmids or viruses, and represent relatively small segments of DNA that contain necessary genetic signals for replication, as well as additional elements for convenience in inserting foreign DNA, identifying cells that contain recombinant DNA, and, where appropriate, expressing the foreign DNA.

The choice of vector for molecular cloning depends on the choice of host organism, the size of the DNA to be cloned, and whether and how the foreign DNA is to be expressed. The DNA segments can be combined by using a variety of methods, such as restriction enzyme/ligase cloning or Gibson assembly.

In standard cloning protocols, the cloning of any DNA fragment essentially involves seven steps: (1) Choice of host organism and cloning vector, (2) Preparation of vector DNA, (3) Preparation of DNA to be cloned, (4) Creation of recombinant DNA, (5) Introduction of recombinant DNA into the host organism, (6) Selection of organisms containing recombinant DNA, (7) Screening for clones with desired DNA inserts and biological properties. These steps are described in some detail in a related article (molecular cloning).

Expression of Recombinant DNA

Following transplantation into the host organism, the foreign DNA contained within the recombinant DNA construct may or may not be expressed. That is, the DNA may simply be replicated without expression, or it may be transcribed and translated so that a recombinant protein is produced. Generally speaking, expression of a foreign gene requires restructuring the gene to include sequences that are required for producing a mRNA molecule that can be used by the host's translational apparatus (e.g. promoter, translational initiation signal, and transcriptional terminator).

Specific changes to the host organism may be made to improve expression of the ectopic gene. In addition, changes may be needed to the coding sequences as well, to optimize translation, make the protein soluble, direct the recombinant protein to the proper cellular or extracellular location, and stabilize the protein from degradation.

Properties of Organisms Containing Recombinant DNA

In most cases, organisms containing recombinant DNA have apparently normal phenotypes. That is, their appearance, behavior and metabolism are usually unchanged, and the only way to demonstrate the presence of recombinant sequences is to examine the DNA itself, typically using a polymerase chain reaction (PCR) test.

If the rDNA sequences encode a gene that is expressed, then the presence of RNA and/or protein products of the recombinant gene can be detected, typically using RT-PCR or western hybridization methods. Gross phenotypic changes are not the norm, unless the recombinant gene has been chosen and modified so as to generate biological activity in the host organism.

Additional phenotypes that are encountered include toxicity to the host organism induced by the recombinant gene product, especially if it is over-expressed or expressed within inappropriate cells or tissues.

In some cases, recombinant DNA can have deleterious effects even if it is not expressed. One mechanism by which this happens is insertional inactivation, in which the rDNA becomes inserted into a host cell's gene. In some cases, researchers use this phenomenon to "knock out" genes in order to determine their biological function and importance.

Another mechanism by which rDNA insertion into chromosomal DNA can affect gene expression is by inappropriate activation of previously unexpressed host cell genes. This can happen, for example, when a recombinant DNA fragment containing an active promoter becomes located next to a previously silent host cell gene, or when a host cell gene that functions to restrain gene expression undergoes insertional inactivation by recombinant DNA.

Recombinant DNA Technology

All gene manipulation is based on microbial genetics—ways of doing in the test tube what bacteria and viruses do naturally. Several of this week's gene manipulations are exemplified by the article on overexpressed angiopoietin in a transgenic mouse.

Gel Electrophoresis

This technique separates molecules on the basis of their size.

- Cast slab of gel material, usually agarose or polyacrylamide. The gel is a matrix of polymers forming sub-microscopic pores.
- The size of the pores can be controlled by varying the chemical composition of the gel.

The gel is set up for electrophoresis in a tank holding pH buffer. Electrodes apply an electric field:

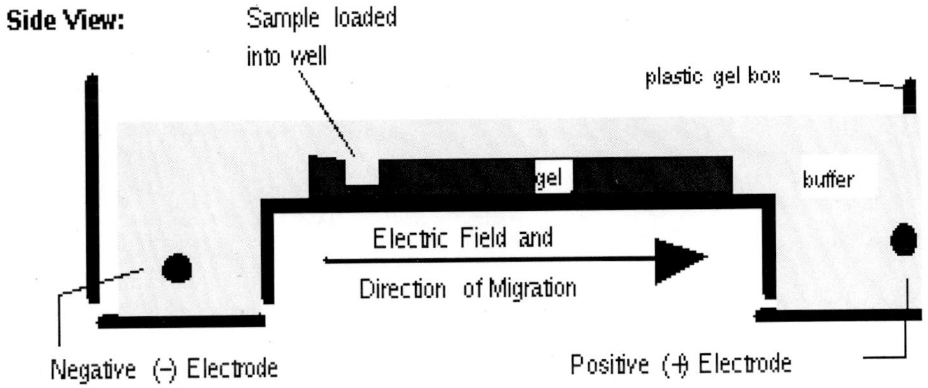

The molecules to separate (DNA RNA) carry a net negative charge (why?) so they move along the electric field toward the positive cathode. (To separate proteins, a detergent would be included which coats the protein with negative charge.) The larger molecules are held up as they try to pass through the pores of the gel, while the smaller molecules are impeded less and move faster. This results in separation by size, with the larger molecules nearer the well and the smaller molecules farther away. Note that this separates on the basis of size (volume in solution), which is not necessarily molecular weight. For example:

- Two DNA molecules of the same molecular weight will run differently if one is supercoiled, because the supercoils constrain the shape to be smaller.
- Two RNA molecules of the same molecular weight will run differently if one has much intramolecular base pairing, making it "smaller."

Aside from the above exceptions, the distance migrated is roughly proportional to the log of the inverse of the molecular weight (the log of 1/MW). Gels are normally depicted as running vertically, with the wells at the top and the direction of migration downwards. This leaves the large molecules at the top and the smaller molecules at the bottom. Molecular weights are measured with different units for DNA, RNA, and protein:

- DNA: Molecular weight is measured in base-pairs, or bp, and commonly in kilobase-pairs (1000bp), or kbp.
- RNA: Molecular weight is measured in nucleotides, or nt, and commonly in kilonucleotides (1000nt), or knt. [Sometimes, bases, or b and kb are used.
- Protein: Molecular weight is measured in Daltons (grams per mole), or Da, and commonly in kiloDaltons (1000Da), or kDa.

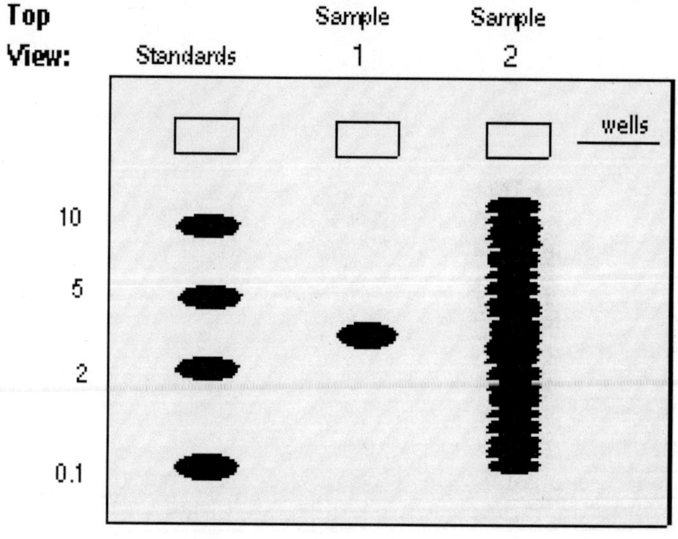

Molecular weight standards run in one well of the gel are used to calibrate the molecular weights of sample molecules. Below is a gel stained with a dye: a colored molecule which binds to a specific class of macromolecules in a sequence-independent manner (probes bind in a sequence-dependent manner). Sample 1 contains only one size class of macromolecule - it could be a plasmid, a pure mRNA transcript, or a purified protein. In this case, you would not have to use a probe to detect the molecule of interest since there is only one type of molecule present. Blotting is usually necessary for samples that are not complex mixtures. By interpolation, its molecular weight is roughly 3.

Sample 2 is what a sample of total DNA cut with a restriction enzyme, total cellular RNA, or total cellular protein would look like in a gel stained with a sequence-independent stain. There are so many bands that it is impossible to find the one we are interested in. Without a probe (which acts like a sequence-dependent stain) we cannot get very much information from a sample like this. Different stains are used for different classes of macromolecules. DNA and RNA are generally stained with ethidium bromide (EtBr), an intercalating agent. The DNA-EtBr complex fluoresces under UV light. Protein is stained with Coomassie Blue or Silver Stain.

Cloning Genes

In nature, DNA molecules recombine for various functions — even DNA between different species. But twenty years ago, despite the work of Barbara McClintock and others, the extent of this recombination was not appreciated. DNA was still thought to be the "master molecule," not to be violated by "unnatural" manipulation.

When scientists began to manipulate DNA in the test tube, many scientists feared that disastrous monsters would result, with unspecified dangers to people. In 1977 scientists at the Asilomar Conference proposed sweeping regulation on so-called "recombinant DNA," technologies which recombine DNA from different species in the test tube. Since then, the dangers have appeared to be little more than those of "natural" genetic mixing. But we remain concerned about issues such as:

- Engineering food crops to resist pesticides. The pesticide resistance genes can escape into natural populations of weeds.
- Engineering a human symbiont microbe, such as E. coli, to produce a deadly toxin such as botulin. In theory this could be done, although it's not clear where such an organism would live, or how well it could "compete" with natural flora.
- Societal dilemmas of human cloning. How far shall we use reproductive technology to shape future humans?

Techniques of Recombinant DNA

How do we manipulate these natural processes for biotechnology; for instance, to make a bacterium that produces large quantities of insulin? One approach would be to cut the appropriate gene from human DNA and paste, or splice, it into a vector such as a plasmid or phage DNA. Our "scissors" are the class of enzymes called restriction endonucleases.

Restriction Endonucleases

An "endonuclease" is an enzyme that cuts duplex DNA in the middle, not at an end (for exonuclease). Different species of bacteria have evolved different restriction endonucleases, each to cut foreign DNA that gets into their cells by mistake. To be cut, the DNA has to lack their own pattern of protective methylation. There are well over a hundred restriction enzymes, each cutting in a very precise way a specific base sequence of the DNA molecule. A restriction endonuclease cuts DNA only at a specific site, usually containing 4-6 base pairs. The enzyme has to cut the DNA backbone twice, recognizing the same type of site; therefore, the site "reads" the same way backwards as forwards—a palindrome.

Restriction Enzyme
Action of EcoRI

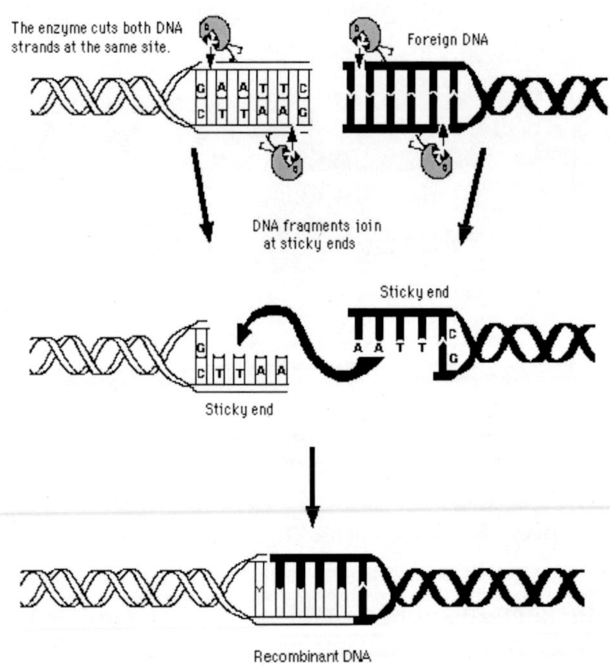

This "sticky ends" from two different DNA molecules can hybridize together; then the nicks are sealed using ligase. (Where does ligase come from? What is its natural function?) The result is recombinant DNA. When this recombinant vector is inserted into E. coli, the cell will be able to process the instructions to assemble the amino acids for insulin production. More importantly, the new instructions are passed along to the next generation of E. coli cells in the process known as gene cloning.

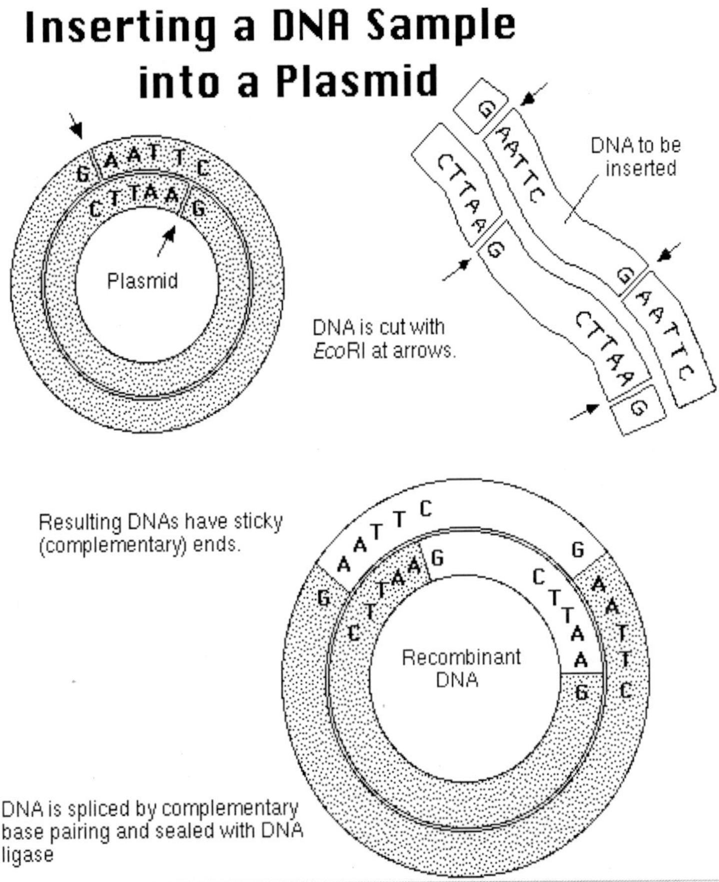

Restriction Site Analysis

How can we use restriction sites to analyze the plasmid products of ligation, and tell whether we in fact have ligated the correct molecule:

Problem: Suggest several "incorrect" ways the plasmid could recombine. Use the PLASMID Program. You MUST practice restriction analysis with this program; it will be on the quiz and/or the test. How do we get the recombinant molecule into a bacterial cell? Usually by transformation (for a plasmid) or by in vitro packaging into a phage head coat (for a phage vector such as lambda phage).

How would we actually locate the appropriately cloned gene? There are many different ways, depending on the specific case. Here is one example, in which a partial sequence of the protein enables us to reverse the code and determine an approximate DNA sequence to use for a radiolabeled probe. The DNA probe will hybridize to clones containing the correct DNA, even if it is just one piece cut out of an entire genome.

Griffiths et al, W. H. Freeman & Co., current edition

The radioactive probe is made by determining a short segment of the protein sequence, then "back translating" to the possible DNA sequences. Short DNA sequences are synthesized to match the protein sequence. Then these DNA oligomers (known as "oligos") are radiolabeled, and applied to the blotted clones. They should hybridize only to clones containing sequence encoding the desired protein.

Reverse Transcription: cDNA Cloning

Suppose we need to clone a gene containing lots of introns. What will happen when the bacterium tries to express it? To overcome this problem, we can start with mRNA isolated from tissues that produce the desired protein. We then use reverse transcriptase enzyme (produced by a retrovirus related to HIV) to reverse transcribe the mRNA into a DNA molecule that now is free of introns. Now we can ligate "sticky ends" onto the cDNA and recombine it into a phage or plasmid vector. Problem: Know the differences between genomic cloning and cDNA cloning. Explain the relative ADVANTAGES and DISADVANTAGES of each technique—depending on the aim of your research.

Gene Therapy

Transgenic Plants

In PCR, a heat-stable DNA polymerase is used, most commonly Taq Polymerase from the thermophilic microbe Thermus aquaticus. Thomas Brock discovered T. aquaticus from a hot spring at Yellowstone National Park.

More recently, an even more heat-resistant polymerase has been developed from a hyperthermophilic microbe growing at 110 degrees C in hydrothermal vent ecosystems in the deep ocean; it's called "Vent Polymerase."

The Taq Polymerase is put with the DNA to be amplified, plus all four NTPs, plus two primers facing each other, about 200 - 6000 kb apart. (Why do we need primers?) The primers are selected based on the DNA region you want to amplify. The tube is placed in a thermal cycler. DNA gets synthesized from each primer, for about 2 minutes. Then the temperature is raised to 95 degrees C — enough to denature (split apart) the DNA base pairs. But the Taq Polymerase remains intact, because it comes from an organism that evolved to grow at this temperature. Now the temperature is decreased again, and primers again can hybridize to the DNA—both the old AND the newly synthesized strands.

Amplification of target sequence

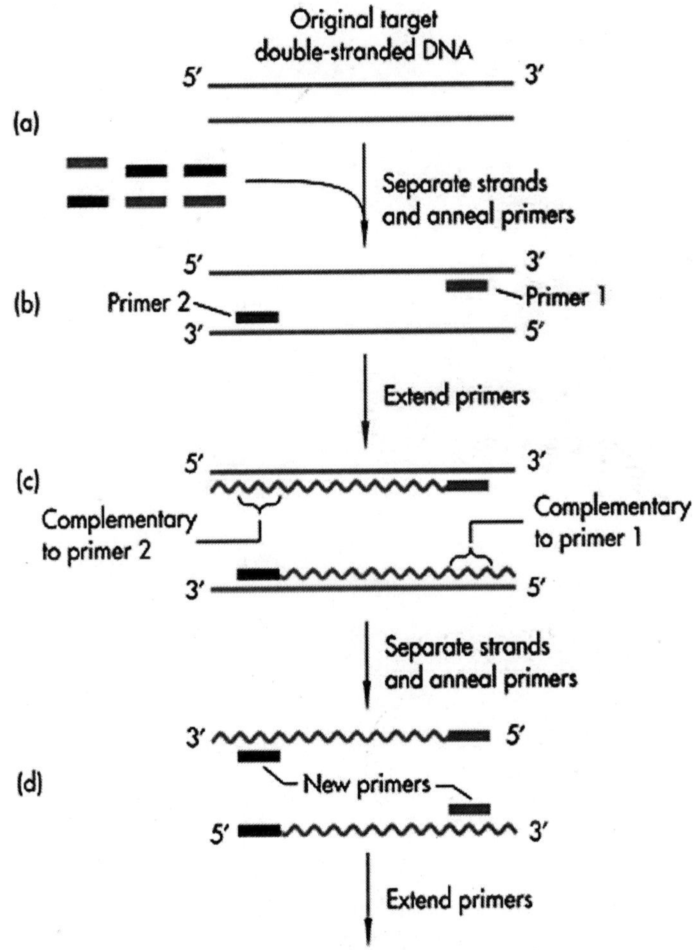

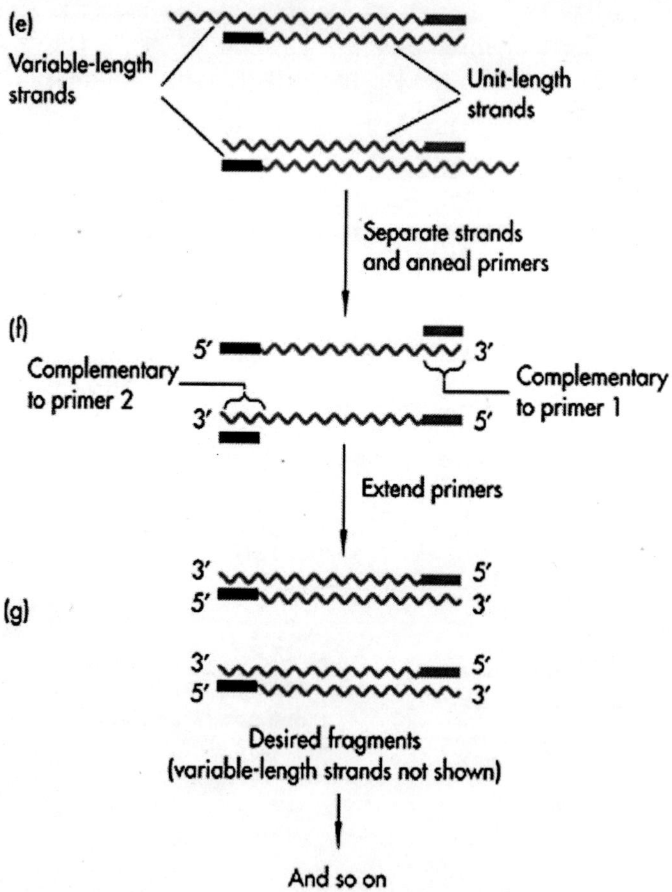

Again, Taq Polymerase extends new DNA strands. Again, the temperature is raised. After repeated cycles, the amount of DNA sequence between the two primers increases exponentially. First 2 strands, then 4, 8, 16, up to about a million. Thus, in a couple of hours, you can get million-fold amplification of a DNA sequence. Griffiths et al, W. H. Freeman & Co., current edition.

Applications of PCR

PCR has replaced cloning for many purposes, particularly the sequencing of DNA. It is faster and requires no vectors, which can mutate as they reproduce. It can be used forensically, to amplify tiny amounts of DNA from criminal evidence; or clinically, to detect DNA sequences linked to inherited disorders.

PCR Experiment

Microbiology Lab: Students characterized the microbial species of various environments in Knox County. Microbial colonies were placed directly into PCR reactions containing primers for amplification of ribosomal RNA genes (rDNA). Lane 4 contains an amplified band of the predicted size, 1000 bp. The top of each well contains genomic DNA. The smears at the bottom contain the PCR primers. The DNA will be sequenced and matched through GenBank to determine the microbial genus.

Thanks to Dan Nickerson '00 and Adam Marks '01.

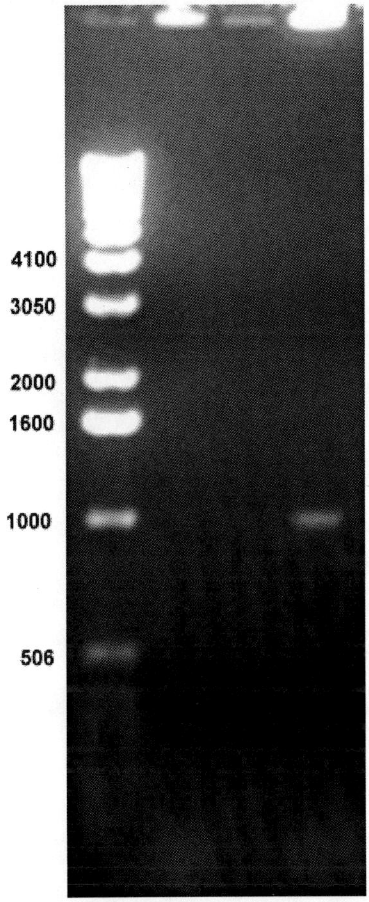

The main limitations of PCR are:

- Only relatively short sequences can be amplified reliably. Anything more than 10,000 base pairs is unlikely to be amplified.
- You need to know the right primer sequences to use, at both ends of the sequence you want to amplify. If two related genes have the same end sequences, you might amplify the wrong gene.
- You only obtain a DNA fragment. To see this DNA at work inside a living organism, some type of cloning has to be done.
- PCR Technology - discussion of preparation of the sample, the master mix, and primers, and detection and analysis of the reaction products.

Gene Sequence Analysis Modified from MIT Hypertextbook.

The sequence of DNA base pairs can be analyzed by

- Restriction mapping. Construct a "road map" of restriction sites. A program to do this is WebCutter.
- DNA Sequence Analysis. Cut and clone various restriction fragments, and determine the exact sequence of base pairs. All sequence information is deposited in GenBank. If you just want the sequence of the peptide translated from the RNA, you have to look for insulin mRNA or cDNA.

Once we have a piece of DNA cloned, it is amplified (available in many copies) and we now have a living clone which provides, in theory, an indefinite source of the DNA sequence. How do we analyze the sequence?

DNA Sequence Determination

Once you have identified a particular region of DNA of interest, you need to find out the precise sequence of DNA nucleotides. This is done by di-deoxy sequencing, in which a DNA polymerase is put together with dNTPs in four different reactions, each containing a small amount of one di-deoxy NTP (ATP, TTP, CTP, or GTP). The di-deoxy nucleotide lacks a 3'OH to continue chain extension, so the chain terminates. Each reaction produces a population of fragments terminated at A, T, C, or G.

- The fragments are either radiolabeled or enzymatically labeled. They can be separated on a gel, or on a fluorescence analyzer. All published DNA sequences in the world are deposited in GenBank.

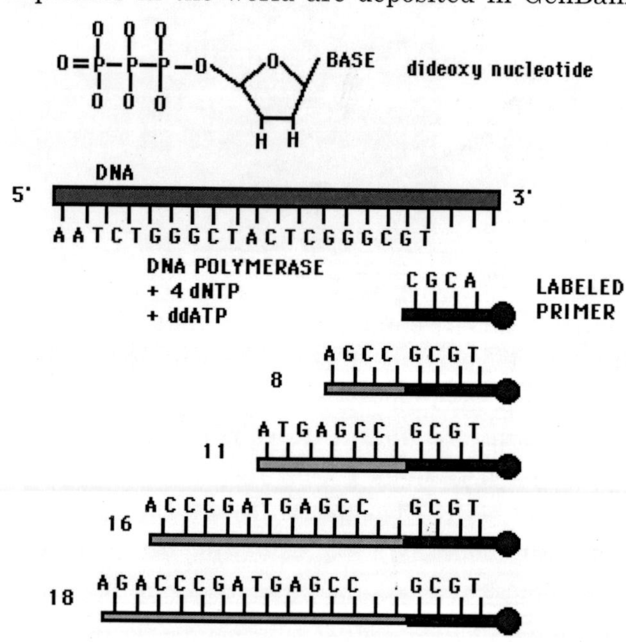

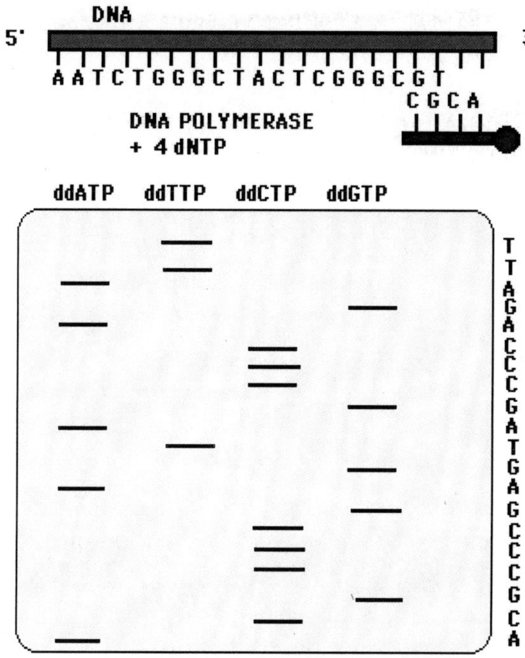

Historically, the products of dideoxy sequencing reactions were subjected to electrophoresis in four different lanes of a gel. Each lane contained a reaction using a different dideoxy terminator nucleotide. The data looked like the image at right.

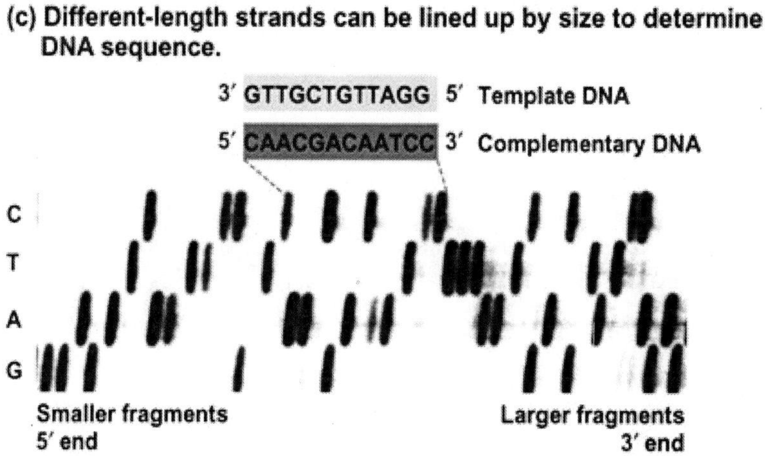

Complementarity and Hybridization

How do we find genes of interest in a genome? Or a particular mRNA in the total cellular RNA? Or a particular protein out of all cell protein?

In solution, hybrid molecular complexes (usually called hybrids) of the following types can exist (other combinations are possible):

- DNA-DNA. A single-stranded DNA molecule (ssDNA probe) can form a double-stranded, base-paired hybrid with a ssDNA target if the probe sequence is the reverse complement of the target sequence. A radiolabeled DNA probe can be applied to DNA from a gel transferred to a membrane, called a Southern Blot (named for its inventor).
- DNA-RNA. A single-stranded DNA (ssDNA) probe molecule can form a double-stranded, base-paired hybrid with an RNA (RNA is usually a single-strand) target if the probe sequence is the reverse complement of the target sequence. An RNA can be radiolabeled to probe a Southern Blot; or, a ssDNA probe can be applied to membrane-bound RNA, called a Northern Blot (name is a pun on Southern.)
- Protein-Protein. An antibody probe molecule (antibodies are proteins) can form a complex with a target protein molecule if the antibody's antigen-binding site can bind to an epitope (small antigenic region) on the target protein. In this case, the hybrid is called an 'antigen-antibody complex' or 'complex' for short. A radiolabeled antibody can probe membrane-bound proteins, called a Western Blot (an even worse pun.)

There are two important features of hybridization:

- Hybridization reactions are specific - the probes will only bind to targets with complimentary sequence (or, in the case of antibodies, sites with the correct 3-d shape).
- Hybridization reactions will occur in the presence of large quantities of molecules similar but not identical to the target. That is, a probe can find one molecule of target in a mixture of zillions of related but non-complementary molecules.

These properties allow you to use hybridization to perform a molecular search for one DNA molecule, or one RNA molecule, or one protein molecule in a complex mixture containing many similar molecules.

Southern, Northern, and Western Blots. Link to summary.

Blots are named for the target molecule.

Southern Blot—DNA cut with restriction enzymes - probed with radioactive DNA. Northern Blot—RNA - probed with radioactive DNA or RNA.

Example—used to measure angiopoietin angiopoietin expression from cDNA in transgenic mouse.

Western Blot—Protein - probed with radioactive or enzymatically-tagged antibodies.

These molecules must then be immobilized on a solid support, so that they will remain in position during probing and washing. The probe is then added, the non-specifically bound probe is removed, and the probe is detected. The place where the probe is detected corresponds to the location of the immobilized target molecule. This process is diagrammed below:

Preparing for Blots

- Southern Blots. DNA is first cut with restriction enzymes and the resulting double-stranded DNA fragments have an extended rod conformation without pre-treatment.
- Northern Blots. Although RNA is single-stranded, RNA molecules often have small regions that can form base-paired secondary structures. To prevent this, the RNA is pre-treated with formaldehyde.
- Western Blots. Proteins have extensive 2' and 3' structures and are not always negatively charged. Proteins are treated with the detergent SDS (sodium dodecyl sulfate) which removes 2' and 3' structure and coats the protein with negative charges.

Transfer to Solid Support. After the DNA, RNA, or protein has been separated by molecular weight, it must be transferred to a solid support before hybridization. (Hybridization does not work well in a gel.) This transfer process is called blotting and is why these hybridization techniques are called blots.

Usually, the solid support is a sheet of nitrocellulose paper (sometimes called a filter because the sheets of nitrocellulose were originally used as filter paper), although other materials are sometimes used. DNA, RNA, and protein stick well to nitrocellulose in a sequence-independent manner.After a series of treatment steps, the probe is added. The probe hybridized to the target molecules is visualized either by autoradiography or by enzyme reaction.

Summary. The important properties of the three blotting procedures of DNA analysis:

Table

	Southern	**Northern**	**Western**
What is separated by molecular weight? (target)	DNA cut with restriction enzymes	RNA denatured with formaldehyde	Protein denatured with SDS
Probe	radioactive gene X DNA	radioactive gene X DNA	Antibody against protein X, labeled with radioactivity or enzyme
What do you learn from it?	Restriction map of gene X in chromosome	-how much gene X mRNA is present? -how long is gene X mRNA?	-how much protein X is present? -how big is protein X?

DNA Microarrays

We can now put most of the protein-encoding genes onto a microarray chip, using technology based on the DNA silicon chip industry. The chip can be used to hybridize to cellular RNA, and measure the expression rates of a large number of genes in a cell.

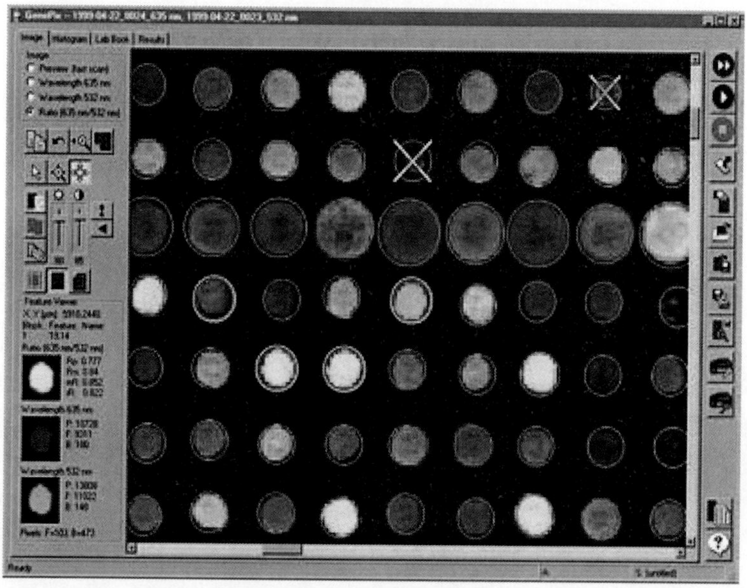

Figure: *Axon Industries.*

Transgenics

Cloning in Animals.

Animals have two different classes of cells: germ line and somatic.

Only alterations in the germ cells can be transmitted to future generations. However, some forms of somatic cell gene therapy can be useful in treating patients. For example, people with cystic fibrosis can receive the cloned CFTR gene in a nasal spray, which then infects the lining of their lungs and improves lung function. The infected (transformed) epithelial cells eventually are lost, however, during normal processes of tissue growth, so the treatment needs to be repeated. Germ-line cloning. There are two basic ways to clone a gene in the mammalian germ line.

- Inject a DNA fragment containing your gene into the nucleus of a fertilized egg (germ-line gene cloning) or of body tissues in a mature host (somatic gene cloning). The DNA gets taken up at random somewhere in one of the chromosomes.

An example of germ-line cloning is the injection of angiopoietin cDNA into the fertilized mouse egg.

(An example of somatic cloning is gene therapy for cystic fibrosis: inhale a vector containing the CF gene, which gets incorporated into the DNA of cells lining the lungs. Somatic cloned genes are not inherited by offspring.)

- Put a transgene containing positive and negative selective markers into an embryonic stem cell tissue culture. The ES cells take up the gene by homologous recombination, replacing the host allele. The ES cells now are injected into a blastula of an unrelated host, and some of the next generation progeny arise from the ES cells. Once you have cloned a valuable gene into an animal, say a sheep that makes insulin in its milk, how can we produce lots of identical progeny quickly without sexual reproduction? This is what the popular press calls "cloning

But when you create a clone, how do you know it worked? You need to know:

- DNA: Did the transgene really incorporate into the genome? Where?
- RNA: Is the RNA expressed? (Or prevented from expression, by a null allele?)
- Protein: Is a desired protein expressed?

Cloning in Plants

Plants don't have separate somatic and germ cells, so creating transgenic plants is easier than creating transgenic animals. Techniques include the use of microprojectile bombardment (the "gene gun") and the use of Agrobacterium tumefaciens.

8

Mechanisms of Cultural Evolution

In DIT, the evolution and maintenance of cultures is described by five major mechanisms, natural selection of cultural variants, random variation, cultural drift, guided variation and transmission bias.

Natural Selection

Cultural differences among individuals can lead to differential survival of individuals, with those that survive more likely to have their cultural variants adopted by others. The patterns of this selective process depend on transmission biases and can result in behavior that is more adaptive to a given environment.

Random Variation

Random variation arises from errors in the learning, display or recall of cultural information.

Cultural Drift

Cultural drift is a process roughly analogous to genetic drift in evolutionary biology. In cultural drift, the frequency of cultural traits in a population may be subject to random fluctuations due to chance variations in which traits are observed and transmitted (sometimes called "sampling error"). These fluctuations might cause cultural variants to disappear from a population. This effect should be especially strong in small populations. A model by Hahn and Bently shows that cultural drift gives a reasonably good approximation for changes in American baby names. Drift processes have also been suggested to explain changes in archaeological pottery and technology patent applications

. Changes in the songs of song birds are also thought to arise from drift processes, where distinct dialects in different groups occur due to errors in songbird singing and acquisition by successive generations. Cultural drift is also observed in an early computer model of cultural evolution.

Guided Variation

Cultural traits may be gained in a population through the process of individual learning. Once an individual learns a novel trait, it can be transmitted to other members of the population. The process of guided variation depends on an adaptive standard that determines what cultural variants are learned.

Biased Transmission

Understanding the different ways that culture traits can be transmitted between individuals has been an important part of DIT research since the 1970s. Transmission biases occur when some cultural variants are favored over others during the process of cultural transmission. Boyd and Richerson (1985) defined and analytically modeled a number of possible transmission biases. The list of biases has been refined over the years, especially by Henrich and McElreath.

Content Bias

Content biases result from situations where some aspect of a cultural variant's content makes them more likely to be adopted. Content biases can result from genetic preferences, preferences determined by existing cultural traits, or a combination of the two. For example, food preferences can result from genetic preferences for sugary or fatty foods and socially-learned eating practices and taboos. Content biases are sometimes called "direct biases."

Context Bias

Context biases result from individuals using clues about the social structure of their population to determine what cultural variants to adopt. This determination is made without reference to the content of the variant. There are two major categories of context biases: (1) model-based biases, and (2) frequency-dependent biases.

Model-based Biases

Model-based biases result when an individual is biased to choose a particular "cultural model" to imitate. There are four major categories of model-based biases: (1) prestige bias, (2) skill bias, (3) success bias, (4) similarity bias. A "prestige bias" results when individuals are more likely to imitate cultural models that are seen as having more prestige. A measure of prestige could be the amount of deference shown to a potential cultural model by other individuals. A "skill bias" results when individuals can directly observe different cultural models performing a learned skill and are more likely to imitate cultural models that perform better at the specific skill. A "success bias" results from individuals preferentially imitating cultural models that they determine are most generally successful (as opposed to successful at a specific skill as in the skill bias.) A "similarity bias" results when individuals are more likely to imitate cultural models that are perceived as being similar to the individual based on specific traits.

Frequency-dependent Biases

Frequency-dependent biases result when an individual is biased to choose particular cultural variants based on their perceived frequency in the population. The most explored frequency-dependent bias is the "conformity bias." Conformity biases result when individuals attempt to copy the mean or the mode cultural variant in the population. Another possible frequency dependent bias is the "rarity bias." The rarity bias results when individuals preferentially choose cultural variants that are less common in the population. The rarity bias is also sometimes called a "nonconformist bias".

Social Learning and Cumulative Cultural Evolution

In DIT, the evolution of culture is dependent on the evolution of social learning. Analytic models show that social learning becomes evolutionarily beneficial when the environment changes with enough frequency that genetic inheritance can not track the changes, but not fast enough that individual learning is more efficient. While other species have social learning, and thus some level of culture, only humans, some birds and chimpanzees are known to have cumulative culture.

Boyd and Richerson argue that the evolution of cumulative culture depends on observational learning and is uncommon in other species because it is ineffective when it is rare in a population. They propose that the environmental changes occurring in the Pleistocene may have provided the right environmental conditions.

Michael Tomasello argues that cumulative cultural evolution results from a "ratchet effect" that began when humans developed the cognitive architecture to understand others as mental agents. Tomasello proposed in the 80s that there are some disparities between the observational learning mechanisms found in humans and great apes - which go some way to explain the observable difference between great ape traditions and human types of culture.

Cultural Group Selection

Although group selection is commonly thought to be nonexistent or unimportant in genetic evolution, DIT predicts that, due to the nature of cultural inheritance, it may be an important force in cultural evolution. The reason group selection is thought to operate in cultural evolution is because of conformist biases. Conformist biases make it difficult for novel cultural traits to spread through a population. Conformist bias also helps maintain variation between groups.

These two properties, rare in genetic transmission, are necessary for group selection to operate. Based on an earlier model by Cavalli-Sforza and Feldman, Boyd and Richerson show that conformist biases are almost inevitable when traits spread through social learning, implying that group selection is common in cultural evolution.

Analysis of small groups in New Guinea imply that cultural group selection might be a good explanation for slowly changing aspects of social structure, but not for rapidly changing fads. The ability of cultural evolution to maintain intergroup diversity is what allows for the study of cultural phylogenetics.

Historical Development

The idea that human cultures undergo a similar evolutionary process as genetic evolution goes back at least to Darwin In the 1960s, Donald T. Campbell published some of the first theoretical work that adapted principles of evolutionary theory to the evolution of cultures. In 1976, two developments in cultural evolutionary theory set the stage for DIT.

In that year Richard Dawkins's The Selfish Gene introduced ideas of cultural evolution to a popular audience. Although one of the best-selling science books of all time, because of its lack of mathematical rigor, it had little impact on the development of DIT. Also in 1976, geneticists Marcus Feldman and Luigi Luca Cavalli-Sforza published the first dynamic models of gene-culture coevolution. These models were to form the basis for subsequent work on DIT, heralded by the publication of three seminal books in 1980 and 1981.

The first was Charles Lumsden and E.O. Wilson's Genes, Mind and Culture. This book outlined a series of mathematical models of how genetic evolution might favor the selection of cultural traits and how cultural traits might, in turn, affect the speed of genetic evolution. While it was the first book published describing how genes and culture might coevolve, it had relatively little impact on the further development of DIT. Some critics felt that their models depended too heavily on genetic mechanisms at the expense of cultural mechanisms. Controversy surrounding Wilson's sociobiological theories may also have decreased the lasting impact of this book.

The second 1981 book was Cavalli-Sforza and Feldman's Cultural Transmission and Evolution: A Quantitative Approach. Borrowing heavily from population genetics and epidemiology, this book built a mathematical theory concerning the spread of cultural traits.

It describes the evolutionary implications of vertical transmission, passing cultural traits from parents to offspring; oblique transmission, passing cultural traits from any member of an older generation to a younger generation; and horizontal transmission, passing traits between members of the same population. The next significant DIT publication was Robert Boyd and Peter Richerson's 1985 Culture and the Evolutionary Process.

This book presents the now-standard mathematical models the evolution of social learning under different environmental conditions, the population effects of social learning, various forces of selection on cultural learning rules, different forms of biased transmission and their population-level effects, and conflicts between cultural and genetic evolution. The book's conclusion also outlined areas for future research that are still relevant today.

Current and Future Research

In their 1985 book, Boyd and Richerson outlined an agenda for future DIT research. This agenda, outlined below, called for the development of both theoretical models and empirical research. DIT has since built a rich tradition of theoretical models over the past two decades. However, there has not been a comparable level of empirical work.

In a 2006 interview Harvard biologist E. O. Wilson expressed disappointment at the little attention afforded to DIT:

"...for some reason I haven't fully fathomed, this most promising frontier of scientific research has attracted very few people and very little effort."

Kevin Laland and Gillian Brown attribute this lack of attention to DIT's heavy reliance on formal modeling.

"In many ways the most complex and potentially rewarding of all approaches, [DIT], with its multiple processes and cerebral onslaught of sigmas and deltas, may appear too abstract to all but the most enthusiastic reader. Until such a time as the theoretical hieroglyphics can be translated into a respectable empirical science most observers will remain immune to its message."

Economist Herbert Gintis disagrees with this critique, citing empirical work as well as more recent work using techniques from behavioral economics. These behavioral economic techniques have been adapted to test predictions of cultural evolutionary models in laboratory settings as well as studying differences in cooperation in fifteen small-scale societies in the field.

Since one of the goals of DIT is to explain the distribution of human cultural traits, ethnographic and ethnologic techniques may also be useful for testing hypothesis stemming from DIT. Although findings from traditional ethnologic studies have been used to buttress DIT arguments, thus far there have been little ethnographic fieldwork designed to explicitly test these hypotheses.

A major difficulty in using existing ethnographic data to test DIT is that traditional ethnographic data tends to ignore individual and intragroup cultural variation and focus almost entirely on intergroup variation.

DIT has been viewed as having great potential for unifying diverse academic fields under one overarching theory. Mesoudi, et al. have identified DIT as the ideal way to build a comprehensive theory of cultural evolution to answer questions about human behavior at different temporal and spacial scales.

Along with game theory, Herb Gintis has named DIT one of the two major conceptual theories with potential for unifying the behavioral sciences, including economics, biology, anthropology, sociology, psychology and political science. Because it addresses both the genetic and cultural components of human inheritance, Gintis sees DIT models as providing the best explanations for the ultimate cause of human behavior and the best paradigm for integrating those disciplines with evolutionary theory. In a review of competing evolutionary perspectives on human behavior, Laland and Brown see DIT as the best candidate for uniting the other evolutionary perspectives under one theoretical umbrella.

Relation to Other Fields

Sociology and Cultural Anthropology

Two major topics of study in both sociology and cultural anthropology are human cultures and cultural variation. However, Dual Inheritance theorists charge that both disciplines too often treat culture as a static superorganic entity that dictates human behavior. Cultures are defined by a suite of common traits shared by a large group of people. DIT theoriests argue that this doesn't sufficiently explain variation in cultural traits at the individual level. By contrast, DIT models human culture at the individual level and views culture as the result of a dynamic evolutionary process at the population level.

Human Sociobiology and Evolutionary Psychology

Human sociobiologists and evolutionary psychologists try to understand how maximizing genetic fitness, in either the modern era or past environments, can explain human behavior. To the human sociobiologist and evolutionary psychologist, culture is either trivial or so bound by genetic fitness that it is unimportant. When faced with a common and seemingly maladaptive trait, practitioners from these disciplines try to determine how the trait actually increases genetic fitness (maybe through kin selection or by speculating about early evolutionary environments). Dual Inheritance theorists, in contrast, will consider a variety of genetic and cultural processes in addition to natural selection on genes.

Human Behavioral Ecology

Human behavioral ecology (HBE) and DIT have a similar relationship to what ecology and evolutionary biology have in the biological sciences. HBE is more concerned about ecological process and DIT more focused on historical process. One difference is that human behavioral ecologists often assume that culture is a system that produces the most adaptive outcome in a given environment. This implies that similar behavioral traditions should be found in similar environments. However, this is not always the case. A study of African cultures showed that cultural history was a better predictor of cultural traits than local ecological conditions.

Memetics

Memetics, which comes from the meme idea described in Dawkins's The Selfish Gene, is similar to DIT in that it treats culture as an evolutionary process that is distinct from genetic transmission. However, there are some philosophical differences between memetics and DIT. One difference is that memetics' focus is on the selection potential of discrete replicators (memes), where DIT allows for transmission of both non-replicators and non-discrete cultural variants. DIT does not assume that replicators are necessary for cumulative adaptive evolution.

DIT also more strongly emphasizes the role of genetic inheritance in shaping the capacity for cultural evolution. But perhaps the biggest difference is a difference in academic lineage. Memetics as a label is more influential in popular culture than in academia. Critics of memetics argue that it is lacking in empirical support or is conceptually ill-founded, and question whether there is hope for the memetic research program succeeding. Proponents point out that many cultural traits are discrete, and that many existing models of cultural inheritance assume discrete cultural units, and hence involve memes.

Criticisms

A number of criticisms of DIT have been put forward.From some points of view, use of the term 'dual inheritance' to refer to both what is transmitted genetically and what is transmitted culturally is technically misleading. Some biologists distinguish between inherited information, which is sequestered (encoded in a set of self-assembly instructions and protected from extensive change), and acquired information, which an organism obtains during its lifetime. This, along with debates about terminology, has generated significant resistance to the concept of cultural or behavioral evolution. Many opponents cite horizontal transmission of ideas to be so "different" from the typical vertical transmission (reproduction) in genetic evolution that it is not evolution.

However, 1)even genetic evolution uses non-vertical transmission through the environmental alteration of the genome during life by acquired circumstance: epigenetics, and 2) genetic evolution is also affected by direct horizontal transmission between separate species of plants and strains of bacteria: horizontal gene transfer. Other critics argue that there can be no "dual" inheritance without cultural inheritance being "sequestered" by the biotic genome . Evidence for this process is scarce and controversial. Why this is a demand of critics, however, can be considered unclear as it refutes none of the central claims laid down by proponents of DIT.

More serious criticisms of DIT arise from the choice of Darwinian selection as an explanatory framework for culture. Some argue, cultural evolution does not possess the algorithmic structure of a process that can be modeled in a Darwinian framework as characterized by John von Neumann and used by John Holland to design the genetic algorithm. Forcing culture into a Darwinian framework gives a distorted picture of the process for several reasons. First, some argue Darwinian selection only works as an explanatory framework when variation is randomly generated.

To the extent that transmission biases are operative in culture, they mitigate the effect of Darwinian change, i.e. change in the distribution of variants over generations of exposure to selective pressures. Second, since acquired change can accumulate orders of magnitude faster than inherited change, if it is not getting regularly discarded each generation, it quickly overwhelms the population-level mechanism of change identified by Darwin; it 'swamps the phylogenetic signal'.

However, DIT proponents might reply that, 1) biotic evolution does not function only on randomly generated phenotypes either, since the phenotypes present in a population are the combined result of random and selective effects during the last generation; and 2)that transmission bias would quite often also reinforce "Darwinian change" since it is widely evidenced that Culture has adaptive value in increasing human fitness.

Another discord in opinion stems from DIT opponents' assertion that there exists some "creative force" that is applied to each idea as it is received and before it is passed on, and that this agency is so powerful that it can be stronger than the selective system of other individuals assessing what to teach and whether your idea has merit . But if this criticism was valid then it would be comparatively much easier to argue an unpopular or incorrect concepts than it actually is. In addition, nothing about DIT runs counter to the idea that an internally selective process (some would call creativity) also determines the fitness of ideas received and sent.

In fact this decision making is a large part of the territory embraced by DIT proponents but is poorly understood due to limitations in neurobiology. Related criticisms of the effort to frame culture in Darwinian terms have been leveled by Richard Lewontin , Niles Eldredge, and Stuart Kauffman.

Know Your Meme

Know Your Meme is a web series originally from Rocketboom which documents various Internet memes and other online phenomena, and also investigates new and changing memes through research and the scientific method. The website was acquired in March of 2011 by Cheezburger Network.

Website

The website, named Know Your Meme: Internet Meme Database, currently contains more than 500 confirmed meme entries. The Know Your Meme project started in December 2007 as a limited-run video series created by Kenyatta Cheese, Elspeth Rountree, Jamie Wilkinson, and Andrew Baron. In a manner similar to Wikipedia, anybody with an account can submit meme entries to the website and submit relevant images that help further document the memes.

The administrators have say over what gets confirmed and what gets "Deadpooled" or rejected. Some of the meme entries are graphic and Not Safe For Work. NSFW entries are given warnings. These warning may differ from consequences, such as bans. Know Your Meme also has a forum section, along with a blog and a shop.

The Know Your Meme web site and web series were acquired in March of 2011 by Cheezburger Network for an undisclosed seven-figure amount.

Memetic Engineering

Memetic engineering is a term developed and coined by Leveious Rolando, John Sokol, and Gibran Burchett while they researched and observed the behavior of people after being purposely exposed (knowingly and unknowingly) to certain memetic themes. The term is based on Richard Dawkins' theory of memes.

- The process of developing memes, through meme-splicing and memetic synthesis, with the intent of altering the behavior of others in society or humanity.

- The process of creating and developing theories or ideologies based on an analytical study of societies, cultures, their ways of thinking and the evolution of their minds.
- The process of modifying human beliefs, thought patterns, etc.

Definition

According to the theory , the effect a meme has on society is based on the application of the meme after understanding the qualities essential to the meme. For example, Rolando, Burchett and Sokol expand on their concept and explain that "Race" and "Racism" are memes incorporating several other memes, some of which have positive connotations in societies that reject racism.

According to the theory Memetic Engineering is, simply put, the analysis of an individual or individual's behavior, the selection of specific memes and the distribution or propagation of those memes with the intent of altering the behavior of others. A memetic engineer doesn't particularly have to consciously make the decision to alter another individuals behavior. It can happen unconsciously when specific behavior is observed, transmitted and then replicated within the observer.

The process of creating and developing theories or ideologies based on an analytical study of societies, cultures, their ways of thinking and the evolution of their minds. Memes themselves are neither good nor bad. For example Race is an ideology that is made up of several memes. When a Meme is introduced, those concepts begin to take on their own process of evolution based on the person who adopts the ideology, internalizes it, and reintroduces it into society causing it to spread like a virus.

According to the above theory, typical memetic engineers include scientists, engineers, industrial designers, ad-men, artists, publicists, political activists, and religious missionaries. Dawkins agrees that much of theology and other theoretical aspects of religion can be viewed as the careful, even worshipful, handling of extremely powerful memeplexes with very odd or difficult traits.

Origins of Memetic Engineering

Memetic Engineering developed from diverse influences, including cutting-edge physics of consciousness and memetics research, chaos theory, semiotics, culture jamming, military information warfare, and the viral texts of iconoclasts William S. Burroughs, J.G. Ballard, and Genesis P-Orridge. It draws upon Third Culture sciences and conceptual worldviews for Social Engineering, Values Systems Alignment, and Culture Jamming purposes. An important example of macro-historical memetic engineering analysis explaining how domination, patriarchy, war and violence are culturally programmed is Riane

Eisler's The Chalice and the Blade (San Francisco: Harper SanFrancisco, 1988), which outlines her Dominator and Partnership Culture thesis. The savvy memetic engineer is able to isolate, study, and subtly manipulate the underlying values systems, symbolic balance and primal atavisms that unconsciously influence the individual psyche and collective identity.

A highly educated but susceptible intelligentsia, worldwide travel, and information vectors like the Internet, cable television, and tabloid media, means that hysterical epidemics and disinformation campaigns may become more common.

This warfare will be conducted using aesthetics, symbols, and doctrines as camouflage that will ultimately influence our cultural meme pool. These contemporary life conditions (Historic Times; Geographic Place; Existential Problems; and Societal Circumstances) are explored in books like Carl Sagan's The Demon Haunted World: Science As a Candle in the Dark (New York: Ballantine Books, 1996), John Brockman's The Third Culture: Beyond the Scientific Revolution (New York: Touchstone Books, 1996), and Michael Shermer's Why People Believe Weird Things: Pseudo-science, Superstition, and Other Confusions of Our Time (New York: W.H.

Freeman & Co, 1996). Fictional descriptions of memetic engineering include Isaac Asimov's seminal Foundation Trilogy (New York: Bantam Books, 1991), George Gurdjieff's artificial mythology Beelzebub's Tales to His Grandson (New York: Penguin USA, 1999); Neil Stephenson's novels Snow Crash (New York: Bantam Spectra, 1993) and The Diamond Age (New York: Bantam Spectra, 1996); and Robert W. Chambers' unearthly The King in Yellow (Buccaneer Books, 1996) tome, which influenced seminal horror author H. P. Lovecraft.

Applied Memetic Engineering

Memetic engineering as a social science lends examples of itself in multiple areas and disciplines. It is currently being examined and researched by the US military as a means to counter insurgency and combat terrorism as explained below in From the Clash to the Confluence of Civilizations by Thomas P.M. Barnett an American military geostrategist, the Pentagon and Richard J. Pech's Inhibiting Imitative Terrorism Through Memetic Engineering. Other examples of applied memetic engineering are present but not exclusive to the marketing and advertising industries. The question is whether these individuals can be truly considered memetic engineers. Marketing and advertising professionals create memes on an ongoing basis however, this alone doesn't necessarily qualify them as being memetic engineers.

Few if any actually fall into this category. This is possibly due to a lack of understanding of the various memes that have taken root in their target audience minds. According to the definition, industrial designers, music artist, athletes and other entertainers would more likely better serve this definition.

This is because of their ability to create products, phrases and ideas that disseminate the population triggering a response within the brain causing a cultural phenomena. Game theory provides an empirical means of advancing the science of memetics.

Memetic game theory, attempts to mathematically capture behavior in strategic situations; where an individual's success in making choices depends on the choices of others, based on past experiences, emotional behavior and learned behavior. It also offers a scientific approach to analyzing social interactions.

According to G.Burchett and his research from G. Edward Griffin's book The Creature from Jekyll Island: A Second Look at the Federal Reserve, the concept of currency or money is considered a meme developed by early bankers. Here we have an example of a meme that was created to better serve the economic foundation of a society.

This is a positive example of memetic engineering. However once the meme of legal tender was created it forced individuals to accept and use the currency under the guise of law. Once an individual is forced to accept currency by the government, essentially the government now controls the means of that individual's survival. In the case of the United States legal tender; the economic control of the citizens was placed in hands of the Federal Reserve which is a private entity.

Examples

An example of an engineered meme is Godwin's Law, a meme which propagates on mail-lists, and which its author professes to have initiated to reduce spam on those lists; one version is "When someone posts a metaphor about Nazis the thread is no longer useful."

Richard Pech discusses the concept of memetic engineering within the context of countering mind contagions associated with terrorism. School shootings, for example, may be explained as an attempt to demonstrate the ultimate form of rebellion against a system in which the perpetrators feel ostracised or isolated. Acts of violence might appeal to their egos and the means for achieving this is replicated via the shooting meme.

To re-engineer such a meme and its ability to infest susceptible minds, all information concerning such violence must be portrayed in an unappealing manner. For example, no one wants to be associated with acts of cowardice. By strongly suggesting that such violence is cowardly and the work of disturbed minds, it has less appeal for replication. In this manner the shooting meme has been re-engineered by removing its attraction, and therefore removing its ability to replicate.

Psycholinguistics

Psycholinguistics or psychology of language is the study of the psychological and neurobiological factors that enable humans to acquire, use, comprehend and produce language. Initial forays into psycholinguistics were largely philosophical ventures, due mainly to a lack of cohesive data on how the human brain functioned. Modern research makes use of biology, neuroscience, cognitive science, linguistics, and information theory to study how the brain processes language. There are a number of subdisciplines with non-invasive techniques for studying the neurological workings of the brain; for example, neurolinguistics has become a field in its own right.

Psycholinguistics covers the cognitive processes that make it possible to generate a grammatical and meaningful sentence out of vocabulary and grammatical structures, as well as the processes that make it possible to understand utterances, words, text, etc. Developmental psycholinguistics studies children's ability to learn language.

Areas of study

Psycholinguistics is an interdisciplinary field. Hence, it is studied by researchers from a variety of different backgrounds, such as psychology, cognitive science, linguistics, and speech and language pathology. Psycholinguists study many different topics, but these topics can generally be divided into answering the following questions: (1) how do children acquire language (language acquisition)?; (2) how do people process and comprehend language (language comprehension)?; (3) how do people produce language (language production)?; and (4) how do adults acquire a new language (second language acquisition)? Subdivisions in psycholinguistics are also made based on the different components that make up human language.

Linguistics-related areas:

- Phonetics and phonology are concerned with the study of speech sounds. Within psycholinguistics, research focuses on how the brain processes and understands these sounds.
- Morphology is the study of word structures, especially the relationships between related words (such as dog and dogs) and the formation of words based on rules (such as plural formation).

- Syntax is the study of the patterns which dictate how words are combined to form sentences.
- Semantics deals with the meaning of words and sentences. Where syntax is concerned with the formal structure of sentences, semantics deals with the actual meaning of sentences.
- Pragmatics is concerned with the role of context in the interpretation of meaning.

A researcher interested in language comprehension may study word recognition during reading to examine the processes involved in the extraction of orthographic, morphological, phonological, and semantic information from patterns in printed text. A researcher interested in language production might study how words are prepared to be spoken starting from the conceptual or semantic level. Developmental psycholinguistics study infants' and children's ability to learn and process language.

Theories

Language Acquisition

There are essentially two schools of thought as to how children acquire or learn language, and there is still much debate as to which theory is the correct one. The first theory states that all language must be learned by the child. The second view states that the abstract system of language cannot be learned, but that humans possess an innate language faculty, or an access to what has been called universal grammar.

The view that language must be learned was especially popular before 1960 and is well represented by the mentalistic theories of Jean Piaget and the empiricist Rudolf Carnap. Likewise, the school of psychology known as behaviorism by B.F. Skinner) puts forth the point of view that language is a behavior shaped by conditioned response, hence it is learned.

The innatist perspective began with Noam Chomsky's highly critical review of Skinner's book in 1959. This review helped to start what has been termed "the cognitive revolution" in psychology. Chomsky posited humans possess a special, innate ability for language and that complex syntactic features, such as recursion, are "hard-wired" in the brain. These abilities are thought to be beyond the grasp of the most intelligent and social non-humans. According to Chomsky, children acquiring a language have a vast search space to explore among all possible human grammars, yet at the time there was no evidence that children receive sufficient input to learn all the rules of their language.

Hence, there must be some other innate mechanism that endows a language ability to humans. Such a language faculty is, according to the innatist theory, what defines human language and makes it different from even the most sophisticated forms of animal communication.

The field of linguistics and psycholinguistics since then has been defined by reactions to Chomsky, pro and con. The pro view still holds that the human ability to use language (specifically the ability to use recursion) is qualitatively different from any sort of animal ability.

This ability may have resulted from a favorable mutation or from an adaptation of skills evolved for other purposes. The view that language can be learned has had a recent resurgence inspired by emergentism. This view challenges the "innate" view as scientifically unfalsifiable; that is to say, it can't be tested.

With the amount of computer power increasing since the 1980's, researchers have been able to simulate language acquisition using neural network models. These models provide evidence that there may, in fact, be sufficient information contained in the input to learn language, even syntax. If this is true, then an innate mechanism is no longer necessary to explain language acquisition.

Language Comprehension

One question in the realm of language comprehension is how people understand sentences as they read (also known as sentence processing). Experimental research has spawned a number of theories about the architecture and mechanisms of sentence comprehension.

Typically these theories are concerned with what types of information contained in the sentence the reader can use to build meaning, and at what point in reading does that information become available to the reader. Issues such as "modular" versus "interactive" processing have been theoretical divides in the field.

A modular view of sentence processing assumes that the stages involved in reading a sentence function independently in separate modules. These modulates have limited interaction with one another. For example, one influential theory of sentence processing, the garden-path theory, states that syntactic analysis takes place first.

Under this theory as the reader is reading a sentence, he or she creates the simplest structure possible in order to minimize effort and cognitive load. This is done without any input from semantic analysis or context-dependent information. Hence, in the sentence "The evidence examined by the lawyer turned out to be unreliable," by the time the reader gets to the word "examined" he or she has committed to a reading of the sentence in which the evidence is examining something because it is the simplest parse. This commitment is made despite the fact that it results in an implausible situation; we know that experience that evidence can rarely if ever examine something.

Under this "syntax first" theory, semantic information is processed at a later stage. It is only later that the reader will recognize that her or she needs to revise the initial parse into one in which "the evidence" is being examined.

In this example, readers typically recognize their misparse by the time they reach "by the lawyer" and must go back and re-parse the sentence. This reanalysis is costly and contributes to slower reading times.

In contrast to a modular account, an interactive theory of sentence processing, such as a constraint-based lexical approach assumes that all available information contained within a sentence can be processed at any time. Under an interactive account, for example, the semantics of a sentence (such as plausibility) can come into play early on in order to help determine the structure of a sentence.

Hence, in the sentence above, the reader would be able to make use of plausibility information in order to assume that "the evidence" is being examined instead of doing the examining. There are data to support both modular and interactive accounts; which account is the correct one is still up for debate.

Methodologies

Behavioral Tasks

Many of the experiments conducted in psycholinguistics, especially earlier on, are behavioral in nature. In these types of studies, subjects are presented with linguistic stimuli and asked to perform an action. For example, they may be asked to make a judgment about a word (lexical decision), reproduce the stimulus, or name a visually presented word aloud). Reaction times to respond to the stimuli(usually on the order of milliseconds) and proportion of correct responses are the most often employed measures of performance in behavioral tasks. Such experiments often take advantage of priming effects, whereby a "priming" word or phrase appearing in the experiment can speed up the lexical decision for a related "target" word later.

As an example of how behavioral methods can be used in psycholinguistics research, Fischler (1977) investigated word encoding using the lexical decision task. She asked participants to make decisions about whether two strings of letters were English words. Sometimes the strings would be actual English words requiring a "yes" response), and other times they would be nonwords requiring a "no" response. A subset of the licit words were related semantically (e.g., cat-dog) while others were unrelated (e.g., bread-stem). Fischler found that related word pairs were responded to faster when compared to unrelated word pairs. This facilitation suggests that semantic relatedness can facilitate word encoding.

Eye-movements

Recently, eye tracking has been used to study online language processing. Beginning with Rayner (1978) the importance and informativity of eye-movements during reading was established. Later, Tanenhaus et al. (1995) used the visual-world paradigm to study the cognitive processes related to spoken language. Assuming that eye movements are closely linked to the current focus of attention, language processing can be studied by monitoring eye movements while a subject is presented auditorily with linguistic input.

Neuroimaging

Until the recent advent of non-invasive medical techniques, brain surgery was the preferred way for language researchers to discover how language works in the brain. For example, severing the corpus callosum (the bundle of nerves that connects the two hemispheres of the brain) was at one time a treatment for some forms of epilepsy. Researchers could then study the ways in which the comprehension and production of language were affected by such drastic surgery. Where an illness made brain surgery necessary, language researchers had an opportunity to pursue their research.

Newer, non-invasive techniques now include brain imaging by positron emission tomography (PET); functional magnetic resonance imaging (fMRI); event-related potentials (ERPs) in electroencephalography (EEG) and magnetoencephalography (MEG); and transcranial magnetic stimulation (TMS). Brain imaging techniques vary in their spatial and temporal resolutions (fMRI has a resolution of a few thousand neurons per pixel, and ERP has millisecond accuracy). Each type of methodology presents a set of advantages and disadvantages for studying a particular problem in psycholinguistics.

Computational Modeling

Computational modeling—e.g. the DRC model of reading and word recognition proposed by Coltheart and colleagues—is another methodology. It refers to the practice of setting up cognitive models in the form of executable computer programs. Such programs are useful because they require theorists to be explicit in their hypotheses and because they can be used to generate accurate predictions for theoretical models that are so complex that they render discursive analysis unreliable. Another example of computational modeling is McClelland and Elman's TRACE model of speech perception.

Issues and Areas of Research

Psycholinguistics is concerned with the nature of the computations and processes that the brain undergoes to comprehend and produce language. For example, the cohort model seeks to describe how words are retrieved from the mental lexicon when an individual hears or sees linguistic input.

Recent research using new non-invasive imaging techniques seeks to shed light on just where certain language processes occur in the brain.

There are a number of unanswered questions in psycholinguistics, such as whether the human ability to use syntax is based on innate mental structures or emerges from interaction with other humans, and whether some animals can be taught the syntax of human language.

Two other major subfields of psycholinguistics investigate first language acquisition, the process by which infants acquire language, and second language acquisition. In addition, it is much more difficult for adults to acquire second languages than it is for infants to learn their first language (bilingual infants are able to learn both of their native languages easily).

Thus, sensitive periods may exist during which language can be learned readily. A great deal of research in psycholinguistics focuses on how this ability develops and diminishes over time. It also seems to be the case that the more languages one knows, the easier it is to learn more.

The field of aphasiology deals with language deficits that arise because of brain damage. Studies in aphasiology can both offer advances in therapy for individuals suffering from aphasia, and further insight into how the brain processes language.

Spiral Dynamics

Spiral Dynamics is a theory of human development introduced in the 1996 book Spiral Dynamics by Don Beck and Chris Cowan. The book was based on the theory of psychology professor Clare W. Graves. "Spiral Dynamics" is a registered trademark of the National Values Center, Inc.

Overview

Spiral Dynamics argues that human nature is not fixed: humans are able, when forced by life conditions, to adapt to their environment by constructing new, more complex, conceptual models of the world that allow them to handle the new problems. Each new model transcends and includes all previous models. According to Beck and Cowan, these conceptual models are organized around so-called Memes: systems of core values or collective intelligences, applicable to both individuals and entire cultures.

In spiral dynamics, the term Meme refers to a core value system, acting as an organizing principle, which expresses itself through memes (self-propagating ideas, habits, or cultural practices). The prepended and superscripted letter v indicates these are not basic memes but value systems which include them.

The colors act as reminders for the Life Conditions and Mind Capacities of each system and alternate between cool and warm colors as a part of the model. Within the model, individuals and cultures do not fall clearly in any single category (color). Each person/culture embodies a mixture of the value patterns, with varying degrees of intensity in each.

Spiral Dynamics claims not to be a linear or hierarchical model, although this assertion has been contested. According to Spiral Dynamics, there are infinite stages of progress and regression over time dependent upon the life circumstances of the person/culture, which are constantly in flux. Attaining higher stages of development is not synonymous with attaining a 'better' or 'more correct' values system.

All stages co-exist in both healthy and unhealthy states, whereby any stage of development can lead to undesirable outcomes with respect to the health of the human and social environment.

Memes

First tier Meme levels are focused on different themes for existence, and include almost all of the worldviews, cultures, and mental attitudes up to today. New systems build on adaptations of previous levels and seek to solve problems created by living in those earlier ways.

(Attaching concrete examples to these levels of psychological existence is difficult and often misleading because (a) there can be multiple reasons for the same behavior and (b) centralization in a single level regarding all aspects of living is rare.

Some advocates of Spiral Dynamics argue that these are ways of thinking about things, not types of people. However, Spiral Dynamics does explicitly define a category of people labeled "Spiral Wizards" who have attained equilibrium around "higher level" Memes.

Beige

- Summary: Archaic-instinctive—survivalistic/automatic/reflexological
- Time of origin: c. 100,000 BC
- Description: "Express self to meet imperative physiological needs through instincts of Homo sapiens."

Purple

- Summary: Animistic-tribalistic magical-animistic Tribal order
- Time of origin: 50,000 BC
- Description: "Sacrifice to the ways of the elders and customs as one subsumed in group." This is the level of traditional cultures.

Red

- Summary: Egocentric-exploitive power gods/dominionist
- Time of origin: 7000 BC
- Description: "Express self (impulsively) for what self desires without guilt and to avoid shame." Expressed by the mentality of street gangs, Vikings, etc.

Blue

- Summary: Absolutistic-obedience mythic order—purposeful/ authoritarian
- Time of origin: c. 3000 BC
- Description: "Sacrifice self for reward to come through obedience to rightful authority in purposeful Way." Embodied by fundamentalist religions.
- Note: Amber is Ken Wilber's current name for this stage.

Orange

- Summary: Multiplistic-achievist scientific/strategic
- Time of origin: c. 1000 AD on (as early as 600 AD according to Graves and Calhoun)
- Description: "Express self (calculatedly) to reach goals and objectives without rousing the ire of important others." Expressed in the Scientific Revolution and the Industrial Revolution.

Green

- Summary: Relativistic-personalistic—communitarian/egalitarian
- Time of origin: From 1850 on (surged in early 20th century)
- Description: "Sacrifice self-interest now in order to gain acceptance and group harmony." Expressed in 1960s pluralism and systems theory.

Spiral Dynamics theory sees second tier memes as emerging levels that gradually move away from a focus on subsistence-level concerns of the First Tier, and towards a being-level existence. The existence of two different tiers of psychological development was introduced by the founder Graves. Cowan claims it is possible Graves introduced the tier system as a marketing instrument. Up to today there is no research evidence the two tiers exist. Cowan no longer supports the existence of two tiers but claims the only thing now known about the Yellow and Turquoise systems is that they are more complex versions of Orange and Green. The open-ended theory suggests that the levels Coral and beyond are not yet substantially formed and will solidify as a greater portion of society develops towards those memes.

Yellow

- Summary: Systemic-integrative
- Time of origin: 1950s
- Description: "Express self for what self desires, but to avoid harm to others so that all life, not just own life, will benefit."

Turquoise

- Summary: Holistic
- Time of origin: 1970s
- Description: An integrative system which combines an organism's necessary self-interest with the interests of the communities in which it participates. The theory is still forming.

Development

The colour system was added in the 1970s as a graphic element to decorate training materials used by Cowan and Beck. The term Meme was introduced by Beck and Cowan in Spiral Dynamics where the color language replaced original Graves terminology. Graves had used letter pairs to refer to each level and had not considered any connection with memetics.

Beck and Cowan emphasized 'change states' which are part of the Graves theory. They identify landmarks on the transformational path between the levels. Graves' original theory uses a double helix model to show the interrelatedness of an individual's perception of life conditions with their inner neuronal systems, producing a level of psychological existence. This double helix of two interacting forces is referred to as a spiral in 'Spiral Dynamics.

Following the release of their book, Beck and Cowan taught this theory in two courses, SDI and SDII. They ceased their formal working relationship in 1999. Since this break both have continued to offer training courses of various types as well as carrying out consultancy work. Beck spent several years applying the theory of Spiral Dynamics while working with individuals and groups in South Africa involved in the transition from Apartheid for which he received a legislative honor from the state of Texas. Beck has continued to facilitate sessions using the model in both business and politics.

Beck and Ken Wilber (with his integral theory) became interested in each other's work, resulting in Beck developing a branch of spiral dynamics that he calls Spiral Dynamics Integral. This version of the theory uses integral concepts such as the four quadrants. Beck is also associated with the spiritual teacher, Andrew Cohen. This approach continues to be applied and developed in organisational and societal contexts.

Cowan does not subscribe to these developments and promotes a version of the theory which he describes as remaining more faithful to the original research of Clare Graves and extending from it. He continues to use the term 'Spiral Dynamics' to describe his work since he co-created it.

With his partner, Natasha Todorovic, he has undertaken work in integrating Spiral Dynamics with NLP and other models, and in developing corporate strategy and practical applications. He has also suggested and additional second tier meme Coral which deals with the development of neurological capacities Each of the external websites listed below promotes the organization of one of the two co-authors.

Pathologies

According to Don Beck and Ken Wilber, each Meme has both healthy and unhealthy versions. The pathologies are sometimes referred to as being "mean" as in "Mean Green Meme" (MGM) or "Mean Orange Meme" (MOM). As examples, the MOM includes the extremes of capitalism like exploitation, environmental devastation and a general lack of ethics and sensitivity, while the MGM includes performative contradictions like anti-hierarchy, anti-competition, etc.

Cowan disputes that any credible evidence exists for the existence of the "Mean Green Meme" and that it is a misrepresentation of the theory. While he recognizes the problems mentioned above, Cowan argues that they exist in other value systems as well, such that attaching them only to the Green Meme is too simplistic.

He considers the term "mean" to be inappropriate and a theoretical distortion when questions of adaptation or maladaptation, congruence or ineffectiveness are more to the point. Psychopathology potentially exists at all levels and is a different dimension.

Todorovic offers data and analysis leading to the conclusion that the "Mean Green Meme" conjecture is contradicted by the data. Todorovic further concludes that MGM is a "failure of analysis" based on a variety of misconceptions about spiral dynamics and constitutes an "alarming misdiagnosis" and brands MGM a harmful "form of spiral fundamentalism."

Criticisms of Spiral Dynamics

Critics point out that the model's implications are political as well as developmental and that while the terminology of the theory is self-consciously inclusive, the practical implications of the model can be seen as socially elitist and authoritarian. In his work on the subject, Beck emphasizes that one of the characteristics of "tier two" individuals, also called "Spiral Wizards", is their ability to make superior decisions for all parties concerned and to manufacture consent for their approaches at lower levels using resonant terms and ideas.

In addition to outlining an underlying developmental theory, Spiral Dynamics gives explicit suggestions to these "Wizards" for both consensual and non-consensual management of "lower-tier" individuals.

One critic of Spiral Dynamics, Michel Bauwens, has argued that some conceptions of what it means to be "second tier" have come to resemble Nietzsche's idea of the Übermensch. Co author Cowan has publicly dissociated himself from the ideas that are currently being promoted by his ex-partner Beck in conjunction with Ken Wilber.

Predictive Medicine

Predictive medicine is a rapidly emerging field of medicine that entails predicting disease and instituting preventive measures in order to either prevent the disease altogether or significantly decrease its impact upon the patient (such as by preventing mortality or limiting morbidity). While different prediction methodologies exist, such as genomics, proteomics, and cytomics, the most fundamental way to predict future disease is based on genetics.

Although proteomics and cytomics allow for the early detection of disease, much of the time those detect biological markers that exist because a disease process has already started.

However, comprehensive genetic testing (such as through the use of DNA arrays or full genome sequencing) allows for the estimation of disease risk years to decades before any disease even exists, or even whether a healthy fetus is at higher risk for developing a disease in adolescence or adulthood. Individuals who are more susceptible to disease in the future can be offered lifestyle advice or medication with the aim of preventing the predicted illness.

Current genetic testing guidelines supported by the health care professionals discourage purely predictive genetic testing of minors until they are competent to understand the relevancy of genetic screening so as to allow them to participate in the decision about whether or not it is appropriate for them.

Genetic screening of newborns and children in the field of predictive medicine is deemed appropriate if there is a compelling clinical reason to do so, such as the availability of prevention or treatment as a child that would prevent future disease.

The Goal

The goal of predictive medicine is to predict future disease so that health care professionals and the patient themselves can be proactive in instituting lifestyle modifications and increased physician surveillance, such as bi-annual full body skin exams by a dermatologist or internist if their patient is found to have an increased risk of melanoma or an EKG and cardiology examination by a cardiologist if a patient is found to be at increased risk for a cardiac arrhythmia.

Predictive medicine is intended not for patients but more so for healthy individuals, its purpose being to determine whether susceptibility to a particular disease is increased or not.

A tremendous number of association studies have been published in scientific literature that show associations between specific genetic variants in a person's genetic code and a specific disease. Association and correlation studies have found that a female individual with a mutation in the BRCA1 gene has a 65% cumulative risk of breast cancer. Genetic variants in the Factor V gene is associated with an increased tendency to form blood clots, such as deep vein thrombosis (DVTs).

Genetics tests are expected to reach the market more quickly than new medicines. Myriad Genetics is already generating revenue from genetic tests for BRCA1 and BRCA2. Predictive medicine changes the fundamental paradigm of medicine from being reactive to being proactive and has the potential to significantly decrease the incidence and prevalence of both common and rare diseases.

Examples of Predictive Medicine

Available types of predictive medicine through health care professionals include:

- Newborn screening: Newborn screening is conducted just after birth to identify genetic disorders that can be treated early in life. This testing of infants for certain disorders is one of the most widespread uses of genetic screening - all US states currently test infants for phenylketonuria and congenital hypothyroidism.

  US state law mandates collecting a sample by pricking the heel of a newborn baby to obtain enough blood to fill a few circles on filter paper labeled with names of infant, parent, hospital, and primary physician.

- Diagnostic testing: Diagnostic testing is conducted to aid in the specificity diagnosis or detection of a disease. It is often used to confirm a particular diagnosis when a certain condition is suspected based on the subject's mutations and physical symptoms.

  The diversity in diagnostic testing ranges from common consulting room tests such as measuring blood pressure and urine tests to more invasive protocols such as biopsies.

- Medical bioinformatics: Medical bioinformatics involves determining individual cell molecular parameters by cytomics and single cell-based microarrays.

  A challenge remains to efficiently extract relevant predictive medicine parameters. This information is frequently extracted by computer-assisted identification and characterization of a few cell populations or gene clusters of interest.

- Prenatal testing: Prenatal testing is used to look for diseases and conditions in a fetus or embryo before it is born. This type of testing is offered for couples who have an increased risk of having a baby with a genetic or chromosomal disorder.

  Screening can determine the sex of the fetus. Prenatal testing can help a couple decide whether to abort the pregnancy. Like diagnostic testing, prenatal testing can be noninvasive or invasive.

  Non-invasive techniques include examinations of the woman's womb through ultrasonography or maternal serum screens.

  These non-invasive techniques can evaluate risk of a condition, but cannot determine with certainty if the fetus has a condition. More invasive prenatal methods are slightly more risky for the fetus and involve needles or probes being inserted into the placenta or chorionic villus sampling.

- Carrier testing: Carrier testing is done to identify people who carry one copy of a gene mutation that, when present in both copies, causes a genetic disorder.

  This type of testing is offered to individuals who have genetic disorder in their family history or to people in ethnic groups with increased risk of certain genetic diseases. If both parents are tested, carrier testing can provide information about a couple's risk of having a child with a genetic disorder.

- Preconception testing: Preconception testing is conducted on a person or two potential parents before a child is even conceived. This allows prospective parents to understand the risk of diseases and likelihood of traits in their future offspring.

Companies have already started to apply preconception testing through sperm banks and egg donation services.

A company called Existence Genetics announced in 2010 that it has invented a new patent-pending preconception testing technology that they call the Pythia Approach (it's scientific name is Offspring Projections.

Through the Combined Analysis of Different Individuals, or OP-CADI) that uses a computer system to combine the genetic makeup of two potential parents in-order to predict the risk of a large number of common and rare diseases, as well as the likelihood of specific traits, in future offspring if those two potential parents were to have a child.

Health Benefits

The future of medicine's focus may potentially shift from treating existing diseases, typically late in their progression, to preventing disease before it sets in. Predictive medicine is based on probabilities: while it evaluates susceptibility to diseases, it is not able to predict with 100% certainty that a specific disease will occur.

Unlike many preventive interventions that are directed at groups (e.g., immunization programs), predictive medicine is conducted on an individualized basis. For example, glaucoma is a monogenic disease whose early detection can allow to prevent permanent loss of vision. Predictive medicine is expected to be most effective when applied to polygenic multifactorial disease that are prevalent in industrialized countries, such as diabetes mellitus, hypertension, and myocardial infarction.

With careful usage, predictive medicine methods such as genetic screens can help diagnose inherited genetic disease caused by problems with a single gene (such as cystic fibrosis) and help early treatment.

Some forms of cancer and heart disease are inherited as single-gene diseases and some people in these high-risk families may also benefit from access to genetic tests. As more and more genes associated with increased susceptibility to certain diseases are reported, predictive medicine becomes more useful.

Direct-to-Consumer Genetic Testing

Direct-to-Consumer (DTC) genetic testing enables a consumer to screen his or her own genes without having to go though a health care professional. They can be ordered without the permission of a physician. Variety in DTC tests range from those testing for mutations associated with cystic fibrosis to breast cancer alleles.

DTC tests make the applicability of predictive medicine very real and accessible to consumers. Benefits of DTC testing include this accessibility, privacy of genetic information, and promotion of proactive health care. Risks of obtaining DTC testing are the lack of governmental regulation and the interpreting of genetic information without professional counseling.

Bibliography

Ambrose, E. J., and Easty, D. M.: *Cell Biology: Reading*, Addison- Wesley, Mass, 1970.

Barrington, E. J. W.: *Biochemistry of Primitive Deuterostomians*, London, Academic Press, 1974.

Bauer, M.W., and G. Gaskell : *Biotechnology: The Making of a Global Controversy*, Cambridge: Cambridge University Press, 2002.

Bowler, Peter J.: *The Mendelian Revolution: The Emergence of Hereditarian Concepts in Modern Science and Society*. Baltimore: Johns Hopkins University Press, 1989.

Broach, J.R. : *The Molecular Biology of the Yeast,* Cold Spring Harbor: Cold Spring Harbor Laboratory, 1981.

Carroll SB, Grenier J, Weatherbee SD: *From DNA to Diversity: Molecular Genetics and the Evolution of Animal Design*. Oxford: Blackwell Publishing, 2005.

Chauhan, B.S. : *Principles of Biochemistry and Biophysics,* Laxmi Publications, Delhi, 2008.

Dhruve, J J : *Biochemistry at a Glance*, Agrotech Pub, Delhi, 2007.

Friedberg, E.C. : *DNA Repair,* New York: WH Freeman and Company, 1985.

Ganguly, Smriti : *Biochemistry of Biomolecules,* Pearl Books, Delhi, 2007.

Harris, John : *Wonderwoman and Superman: The Ethics of Human Biotechnology,* Oxford: Oxford University Press, 1992.

Hartl D, Jones E: *Genetics: Analysis of Genes and Genomes*. Jones & Bartlett, 2005.

Kitcher, Philip : *The Lives to Come: The Genetic Revolution and Human Possibilities*, New York: Simon and Schuster, 1996.

Lata Bhattacharya: *Biochemistry of Nutrition*, Discovery, Delhi, 2010.

Lodish, Harvey : *Molecular Cell Biology*, New York: W. H. Freeman, 2000.

Morholt, E. and Brandwein, P.F. : *Sourcebook for the Biological Sciences,* San Diego: Harcourt Brace Jovanovich, 1986.

Morris, Jonathan : *Ethics and Biotechnology*, New York: Chelsea House, 2005.

Narvekar, Raghunath : *Molecular Biochemistry : Principles and Practices,* Adhyayan Pub, Delhi, 2008.

Patel, S V ; U K Kandoliya: *Biochemistry at a Glance*, Agrotech Pub, Delhi, 2007.

Primrose, S. B. .: *Molecular Biotechnology,* Boston: Blackwell Scientific Publications, 1991.

Reuter, Lars : *Modern Biotechnology in Postmodern Times? A Reflection on European Policies and Human Agency,* Dordrecht, Netherlands: Springer, 2003.

Sarath Chandra Bose: *Biochemistry : A Practical Manual*, Pharma Med Press, Delhi, 2010.

Sharma, Pradeep: *Biochemistry and Organisation of Cells*, RBSA Pub, Delhi, 2006.

Singh, Bimla: *Biochemistry and Genomic Revolution*, Vista International Pub, Delhi, 2006.

Trevor H. : *Transforming Matter: A History of Chemistry from Alchemy to the Buckyball,* Baltimore: Johns Hopkins University Press, 2001.

William Alec : *The Chemical Industry*, London: Longman Group Limited, 1971.

Yadav, Neelam: *Biochemistry and Biotechnology : A Laboratory Manual*, Pointer, Delhi, 2007.

Index

H

L

M

N

P

R

S

T

❑❑❑